Beyond God

A scientist's search for the
meaning of life in the
twenty-first century

Beyond God

A scientist's search for the
meaning of life in the
twenty-first century

Emmanuel Papadakis

BOOKS

Winchester, UK
Washington, USA

First published by iff Books, 2013
iff Books is an imprint of John Hunt Publishing Ltd., Laurel House, Station Approach,
Alresford, Hants, SO24 9JH, UK
office1@jhpbooks.net
www.johnhuntpublishing.com
www.iff-books.com

For distributor details and how to order please visit the 'Ordering' section on our website.

Text copyright: Emmanuel Papadakis 2013

ISBN: 978 1 78099 892 3

A CIP catalogue record for this book is available from the British Library.

Design: Stuart Davies

Printed and bound by CPI Group (UK) Ltd, Croydon, CR0 4YY

We operate a distinctive and ethical publishing philosophy in all
areas of our business, from our global network of authors to
production and worldwide distribution.

CONTENTS

Preface 1

Part 1: Acknowledgement 7
Chapter 1
What is the point? 8
What is happiness? 10
Believing nothing 11

Chapter 2
The problem of evolution 14
Consciousness 16
What does science mean? 19

Chapter 3
What is religion? 23
The unbreakable custom 26
Do miracles exist? 28

Chapter 4
Why does evil exist? 31
Why does evil prosper? 33
Overcoming evil 34

Chapter 5
The realm of truth 36
The perception of truth and uncertainty 37
What is the point of philosophy? 40

Chapter 6
Free will: the insoluble paradox? 43
God 46
Death 49

Part 2: Resolution   53
Chapter 7
Resolving to pursue enlightenment   54
The delusion of desire?   55
Dedication   57

Part 3: Pursuance   59
Chapter 8
The law of morality   60
What is right and wrong?   62
The conditions for happiness   65

Chapter 9
The wisdom of Spinoza   67
The problem of ego   68
The key to freedom?   70

Chapter 10
The myth of impulsiveness?   73
The burden of responsibility   74
Discarding anxiety   75

Chapter 11
Passionate illusions   77
Resolving contradictions   78
Understanding and communication   80

Chapter 12
Healthy body, healthy mind?   82
It isn't about whether you win   83
The illusion of intoxication   84

Chapter 13

The use and limitations of intelligence   86
The difference between eloquence and excellence   87
One reality?   88

Chapter 14

The difference between equanimity and indifference   91
The practical use of equanimity   93
An ascetic mind   94

Chapter 15

What is worship?   96
Training the mind   97
Transcending duality and realizing emptiness   99

Chapter 16

Freedom of thought   103
Who's enlightened?   104
Achieving wisdom   106

Chapter 17

Overcoming obstacles   108
Cultivating virtue   109
Criticism and pride   111

Chapter 18

Dedication and expectation   113
Living in the world   114
Realization   116

Notes   121
Glossary   130

Preface

This is not a book about praising the lord or finding your inner self. It is the distillation of a more than decade-long journey to understanding existence and finding a deep inexpressible joy in it by three steps: acknowledging what we have, resolving to realize the purpose for which we exist, and pursuing enlightenment. Acknowledging what we have is realizing that reality is a much stranger mystery than we usually appreciate, science is just a window to this reality not an answer in itself, and that some appreciation of philosophy can help us understand what it is we are and what a rare and wonderful opportunity life presents us with. It also means acknowledging what created us. We can each decide for ourselves if 'God' exists, but either way it means acknowledging life has meaning for each of us if we should choose to realize it. Resolution just means having acknowledged life is worthwhile and has a purpose, we should resolve to pursue it wholeheartedly and without neglect or hesitation. We should resolve to fulfill this purpose, which is to achieve happiness for ourselves and those around us. This doesn't mean happiness in the ordinary sense; it means happiness in the fundamental sense. A deep and valuable fulfillment of existence. Pursuance is how to do this: how to pursue our purpose in the way we live in the world. How should we live and pursue the happiness that is the purpose of every human life?

It actually doesn't matter at all whether you believe in 'God' or not. What matters is whether you appreciate your existence, what you do with it, and how you choose to live.

What is the point of life? The answer seems straightforward: it must be happiness, since if we're not happy what's the point of being alive? But real happiness, a truth more than just temporary pleasure or excitement, is not easy to find. It's also hard to be happy if you think life has no meaning; that it is without purpose. But that's what science seems to suggest. The creation of the earth and evolution appear random, meaningless events in a material universe. Physics and biology have demolished all the strange irrational beliefs about nature and creation that used to exist, sweeping away ideas held for thousands of years. The result is a wide range of views on the point of life; what it's all about. Many think there is no God and life has no ultimate purpose other than to enjoy themselves whilst they're alive. Life is just about enjoying yourself now, before you die: "There is no meaning because god doesn't exist, and when we die, we won't exist anymore either." There's no God and there's nothing more to know. Philosophy has been replaced by science, and anything beyond it is religion which is just a faith in something without proof: ideals we want to believe. Once you realize you're a creature like any other, created like any other by a blind process of evolution, and the driving force of life is survival and reproduction, where is the ultimate meaning?

But this way of thinking is a limited view of both science and philosophy. It's taken for granted because science has shown the nature of reality, there's nothing more to know and believe in. But once you understand some science you can see it's only a door to understanding a world of more profound and vast complexity than mankind ever conceived possible. And as soon as we consider what science really has to say about the nature of reality it's clear there's nothing ordinary about it at all. Likewise philosophy seems irrelevant and displaced by science; by real concrete knowledge of reality. But science is really just a branch of philosophy, not something replacing it.[1] 'Philosophy' is just applying critical reasoned thought in the search for knowledge and understanding. 'Science' is the application of philosophy to the world: the study of what is perceivable to the human senses by observation, idea and experiment. Philosophy is not only able to also consider things beyond science, such as what has meaning and how we should live. It is also able to consider the nature and limitations of perception and knowledge, from which science itself derives. And as any brief study of philosophy will show, whatever 'ultimate' reality may be, the only thing we can be sure of is we know very little about it. Most scientists would add we know very little about the reality we see and touch either.

Because religion is so important in people's lives we often think it must be absolutely true or not at all. But even a short study of the history of any religion shows that religion is not fixed. It has evolved throughout history in thousands of scriptures, teachings and sects as a way of expressing various spiritual and metaphysical[2] beliefs that have evolved over time and depended on interpretation. The fact is that reason and faith need each other. Philosophy shows that we must have some faith in the usefulness of our reason; but we must also have reasonable faith, since irrational faith in any religion just leads to confused and contradictory beliefs. We can't have faith in things that reason and experience contradict, and we should guide our faith by those things they indicate are likely to be true.

We might say the most important thing is love, so we should create it and look after it as much as we're able and that's the meaning of life, nothing more. But whilst this is good it's not enough because there's nothing transcendental about this view, whereas in reality there is something transcendental about existence and its nature; about the fact we exist. The purpose of life is happiness, and love is essential to it, but it's not just love or happiness in an ordinary kind of way. In order to get there you have to think and act: to pursue some understanding of the truth and actually live according to it. This doesn't need great wisdom, obscure practices, or years of living in a monastery; but it does require significant effort from anyone who wants to achieve it. Existence is a strange and phenomenal thing. The meaning of life requires us to both profoundly appreciate it and live our part in it according to the 'right way'. Practicing virtue: living honestly, working hard and sharing with others, is the foundation of this pursuit.

Many people think there's no use in seeking the 'secret of the meaning of life' because it doesn't exist: there is no secret. However such a secret does exist. It's a truth that has run through the heart of all the great religions and the writings of all the great philosophers. The reason it's never been accessible is because its teachings have often been written in subtle and esoteric terms. Yet the simple essential truth of what it has taught is that people should learn to appreciate existence and to improve themselves morally if they want to seek true happiness. It has never appealed for faith based on doctrines alone; rather it has taught that actual experience and the practice of virtue should be the fundamental basis of all true faith. Accordingly, it has traditionally rejected the need for dogmatic and inflexible rules, but not the need for a structured framework to understanding the world and guiding how one chooses to live within it. This means being open-minded but not ill-disciplined; open to change but not to corruption.

Such ancient teachings do not contradict science. In fact, they embrace it, and provide the basis for living according to an ethics that is both humanistic and transcendental; that is concerned with how you live in the world, and at the same time demands each person should seek the ultimate truth for themselves. Life is a gift that should be lived for the sake of happiness, but this happiness requires hard work and struggle just as we must struggle for our own daily existence. This is how we can realize 'the point' for ourselves. Something we are all born with: a happiness and purpose in life undisturbed by life's turbulence yet rooted in existence itself. The only things essential to achieve this are appreciation of what we have and pursuit of virtue. Knowledge helps to bring appreciation: acknowledgement of what we have. Working with discipline towards virtue brings fulfillment: the pursuance of enlightenment. The reason these are difficult to achieve is because self-improvement requires uncomfortable hard work. It's easy to preach about self-awareness and self-control from the mountain top: it isn't easy to get there yourself.

Plato told a famous story about awareness. In it, people are likened to ordinarily sitting in a darkened cave and mistaking the shadows of a puppet-show on the cave wall for what is real. The philosopher is the one who has freed his mind to step out of the cave into the sunlight to see things as they really are; to stop mistaking shadows for reality. This is what it feels like to wake from an existence in which life seems mundane, frustrating and difficult, to one in which it is still frustrating and difficult, but no longer mundane. Something changes that makes each moment of life inexpressibly joyful. Even when you aren't enjoying yourself you feel free to appreciate the fortune to be alive and be a part of something greater than yourself: that which creates us, binds us, governs us, and gives us the freedom to choose what we become. It is this realization which can help us see that life is hard and short for so many that we should each struggle in our own lives to ensure we live as best we can and help all we come into contact with achieve as much happiness and quality in existence as possible. It's also the pursuit of this realization that can resolve some of life's fundamental paradoxes. How can we stay sane when we know we must die? How can we see pleasure and pain equally without being indifferent to living? How can we control desire when it's desire that drives us to live? How can we be at peace in the midst of uncertainty? And how can we accept our fate whilst also resolving to change it?

The meaning of life: what we think is the meaning of our life, governs all human lives, consciously or unconsciously. There is no proof for God's existence and there never has been; this is why the question of 'faith' arises. The question is then, what do you have faith in? This is a journey from believing the truth to be what we can see and touch, to realizing this is just the façade of something much deeper; a colossal edifice that extends far beyond the power of human sight.

Please refer to the glossary for a guide to some key terms, names and phrases.

Part 1: Acknowledgement

At this time there rose a new faith in reason, freedom, and the brotherhood of all men; the new faith, and, as I believe, the only possible faith, of the open society.

Popper, 'Enemies of the Open Society' ['this time' referring to the birth of democracy in ancient Greece; the open society being the free democracy vs. the closed society: the tribal dictatorship]

Chapter 1

What is the point?

In the beginning life is quite straightforward, but after a while things change. Time passes, you grow older, and you begin to wonder what it's really all about. We are born from where we know not, live for a short time filled with much difficulty, and must work most of our waking hours just to earn a living. We eat, sleep, reproduce and chase our desires like everyone else, and then we grow old, decrepit, and we die. A death that is an end in nothingness; we are no more than a decaying corpse, absent from the world and existence. What is the point? Gone in a moment: the time that passes cannot be regained, the people that are gone cannot be recalled. We struggle to live and then pass away. As it is said in Ecclesiastes,

> Everything is meaningless. What profit has a man for all his efforts? One generation passes away, another comes, but nothing changes. The sun rises and goes down to come back up again from where it came. The wind flies to the south and the north, twisting continually, but gets nowhere. All the rivers run into the sea, but the sea is never full; then the water returns from where it came. Desires have no end: the eye of man is never satisfied with seeing, nor the ear with hearing. There is nothing new: what has been done before will be done again; whatever will be done in future has been done already.

What is the purpose of such an existence?

The answer often given is that the point is to be happy whilst we live. That there's so much suffering and hardship in life that to spend one's brief time unhappy or not living rightly is a great waste, so the point is to enjoy life and live it well. But at first it seems that this isn't enough, since we'll still meet the same end whether we're happy or not. Who cares about being happy if you are still going to grow old and die, and if there is no chance of escape? If you must work to live for a futile existence that will end in extinction and nothingness? Over four thousand years ago the poet of Gilgamesh wrote, "Our days are numbered, our occupations a breath of wind. Where is the man who can clamber to heaven?" There is no man who can clamber to heaven, and if this has always been the fortune and fate of men what is the point; what more is there to know? We are born from nothing, we live, and then we die and return to nothing.

But the real answer is that the point of life is not just to be 'happy' in the ordinary sense. It is to be really happy: to appreciate one's existence and find fulfillment in it. This isn't a happiness that depends on luck, or changes with the circumstances. It doesn't come and go, and it isn't a temporary sensation. It is not enough just to 'feel' happy. True happiness can only derive from a conscious fulfillment that is more than just an idea or a feeling of happiness.

Aristotle observed that happiness is the end of all human activities and aims: everything we pursue is to make us happy, but only happiness is pursued for itself alone. When a person achieves real happiness they find there is no need to question the purpose of life for the answer is discovered. This is the point. People spend endless time arguing whether God exists or life has any purpose, but when you realize that you yourself exist, what this existence is worth, and how you may use it to achieve happiness for yourself and others, what need is there to question whether God exists? The answer becomes clear and the question doesn't concern you any more. But to realize this happiness requires both a change of perspective and a change in oneself, since it's not something one is just born with.

What is happiness?

The difficulty for most has never really been understanding that the purpose of life is to be happy, the difficulty has been finding it, since many people think happiness lies in many different places. Some are inclined to pursue wealth, power and fame, since it appears this is where happiness is best found and this is what we enjoy the most. But whilst we should seek a little of these; of wealth to support ourselves, fame if we want to be successful, and power if we want to have any responsibility, there is a problem if these are seen as ends in themselves. We want things or do things because we think they will bring us the happiness we seek, but if true happiness is possessed only by a person who seeks nothing more than they have, then it will never be found by someone who makes power, money or fame the ends of their life, for these quests are endless: they have no end but themselves and cannot bring peace of mind.

The usual reality of great wealth isn't supreme and luxurious happiness, but dissolution and a life dominated by frivolity and waste that always covets more. Rather than finding it brings complete contentment, its most frequent effect is to make people lose so much perspective on what they need to be happy that the more they accumulate the less they are able to find this happiness, until in the end the only distinction between the poor man who spends his days seeking money and the rich man who already has it, is that one has more possessions, though both are as discontented as each other and neither seem to be able to find what they are looking for. Likewise the usual reality of fame isn't certain and lasting happiness or unparalleled joy, but rather the more famous people are the more insecure, unstable and deluded they tend to become about the value of themselves and others. Similarly, our love of power springs from pride and its delusions of grandeur. From its desire for power and immortality: to have the power to act over others but not be acted on itself. But man is limited to this mortal life like any other creature on earth and cannot transcend the confines of his existence any more than an ant. Wisdom and happiness can only be found in reconciling ourselves to this truth, not in an attempt by the ego to overthrow it by imposing itself on the world or others. We desire power because we think it makes us God-like; that by obtaining it we can somehow transcend mortal limitations. But fulfillment can't be obtained by creating unhappiness or oppression for others, or by attempting to deny the true nature of reality.

The reason that we seek wealth, power and fame is because we think having power over others, the praise of others, and the means to indulge pleasure are the foundation of being happy. But considered properly one can see that this isn't the case at all. In the hands of a wise mind wealth or power can be put to a good use; but they do not bring peace of mind, and we don't find the minds of the wealthy and famous to be a mark of the happiest on earth. Most who obtain them end their existence in the ruin of their body with excess or their mind with vainglory. The life of man is brief. It does not require a great deal to attain happiness while it exists and when it ends it takes nothing with it. It is difficult to think of ourselves no longer existing; that one day we'll cease to exist in the world. But however much power, or wealth, or glory we may acquire we cannot surpass this reality. We can only obtain freedom by accepting and reconciling ourselves to it, by learning to live well within it, and by realizing a purpose for ourselves in it. Indeed, the great matter of existence is answering this question 'what is the point?' for ourselves, since what is realized through our own existence and whether we find happiness depend upon how we answer it.

Believing nothing

It's often said the logical consequence of thinking about death and non-existence must be to conclude that the purpose of life is to have as much pleasure and enjoyment as possible whilst we live, because we could die at any minute and will then cease to exist. And since the best means of having pleasure are through power and the senses, what else is there to pursue and what else should we follow? What further meaning is there to be found? The wise and the foolish perish just the same, the hard-working and lazy perish just the same. And the only reason most don't indulge their desires more is because they lack the means to.

The first answer to this however, is that it's a hopeless and unhappy way to live: to believe that the greatest happiness available is through the indulgence of pleasure and pride, that existence has no real purpose but the dissatisfied and endless indulgence of pleasures and the ego. That life is, in the end, an empty and worthless pursuit. We may be biological organisms and our existence brief, but this doesn't mean it has no purpose or pursuing an empty kind of happiness is the means to achieve fulfillment from it. It is those who think there's no more point to life than to enjoy themselves that end up being least happy. It's also no surprise to find that most who think this way and believe this is the nature of life have a feeling of oppression with reality and frustration at everyday existence, instead of being filled with contentment and joy. There's much to be said for enjoying life and 'living for the moment', but life can't be lived recklessly or without thought and responsibility, and the best way to make use of every moment isn't to enjoy ourselves with whatever pleasure we can find. Reality is also deceptive. Many things which appear to give pleasure or be good turn out to be harmful, and the endless pursuit of pleasure doesn't bring unlimited happiness, but only discontent to body and mind.

It was wisely said "The truth shall set you free," and under-standing the great truth depends only on one's own mind. When one realizes it for oneself circumstances and possessions become immaterial. Of course the essentials are necessary, but happiness doesn't need great wealth, fame or power, or any such thing to achieve; and even if you acquire them they may be more of a hindrance than a help. They have some use if they are well used, but they are not the cause or source of happiness, and if used poorly or sought as ends in themselves they destroy it. Many people make a great pretence of being happy, but if we look closer we find that what they speak of is not happiness at all. To be at the mercy of desire is not compatible with happiness; it is to dissipate life, not fulfill it. And to become ruled by the influence of greed and indulgence is not to attain happiness but to squander it.

If we wish to be happy and perceive the joy of life we must learn to distinguish what is good from what appears so, to attain some self-control, and achieve some balance within ourselves and our relation to the world. Then we can be free from confusion and discontent. Learning to have a real appreciation of life, and realizing the importance of virtue, are the beginning of learning to find true joy and purpose in life and not merely the illusion of them. Everyone has the spirit; the gift of a conscious mind and the power to choose what they become. How far happiness is achieved just depends on realizing this and making use of it.

Chapter 2

The problem of evolution

The problem that faces many people is that even though they realize life isn't worthless and there are some things greater than oneself to be pursued, they also realize that humans are mere products of biological evolution: a blind and accidental process of biology in a chance physical universe. We are descended from a species of single-celled organism and share 97 per cent of our genetic code with the chimpanzee. And as any student of zoology will tell, there often appears little more to nature than a brief and nasty struggle for existence in the aim to survive and reproduce.

Even worse, evolution explains why parents love their children even more than themselves, why men's chief desire is to pursue as many women as possible and why self-interest is the governing principle of all life. 'Love' is just the emotional instinct created to ensure we protect those near to us so that our family genes survive and are carried to the next generation. Whoever carries these genes, and whoever is related to us, it is in our interest to protect so that these genes survive. This is the driving force of all evolution. It's only because our family, and to a lesser degree our friends, carry the same genes as us that we're interested in their welfare. It's why we are so self-interested, and it's why species of life cooperate with each other until it no longer serves their interest to do so.

This has led us to see that in many ways we are no different from other animals, and like other animals our lives are nothing but a brief competition driven by the desires to survive and reproduce. But the answer is so what if nature explains all? This isn't telling us anything we didn't know. It didn't take the discovery of genetics for mankind to realize the driving force of life was reproduction. And just because modern science gives heredity a different name doesn't change anything, since people have always known they're created to think, feel and act in a particular way, and whether you call it 'human nature' or 'genetic programming' makes no difference. Anyone observing earth would have long known that self-interest governs much behaviour, and seen that in the grand scheme the interests of individuals are always subservient to the species; because it's upon the species surviving that individuals depend.

Because nature explains much of how we act and because we share some of it with other animals doesn't mean it's worthless: it just means we should see that we are a part of nature, not something separate from it, and if we are to be happy we should respect the life around us and fulfill our existence as part of it. An 'interest' in its own survival is obvious for anything that wishes to exist, as is a means of creating the next generation. And as for genes, they do not intend selfishness or self-interest and the process of evolution merely happens as it must. Genes evolved because they were a viable way to encode life; humans evolved as genes cooperated in ever more complex organisms. Indeed, the trillions of cells in complex organisms are models of infinite cooperation, as are most of their societies.[3] If self-interest was the only foundation of life there would be none; co-operation of genes, of individuals, and of species, is required to sustain it, and it is only within this context that a necessary competition takes place. Moreover the process of evolution is not deterministic; genes programme and are re-programmed within each individual. People may talk as if we are tools used by genes to perpetuate themselves across generations, but the only truth this conveys is that species last longer than individuals. Genes are not distinct from the individuals that carry them: it is the expression of genes that creates the organism, and it is through each unique individual that each unique set of genes is created.

Consciousness

Above all this, the point in which man is unique and differs from all other life on this earth is that he is self-conscious. He is free to consider and reflect upon his actions and act accordingly. Meaning comes from consciousness. As soon as you know, and you know that you know, you must do what is right; and this is all that matters. An animal can't choose to act other than it does and we don't blame it because we know it must follow its instincts. But the genetic code has evolved to create something special in us: a mind that isn't just pre-programmed but can decide what it wills. A man has the power of understanding and can do as he will: live as he will, and die as he will.

Having studied the origins of life it can be hard to throw off the idea that there's no design or intention to life, just an inexorable and meaningless process of replication through modification and selection: the propagation of genes driven by self-interest and competition. And it may seem difficult to draw a purpose from this, for where are the truth and beauty to be found in the replication of genes, and us as the blind means of reproduction? But whilst biology can explain how life evolved and much of what it is, it cannot explain why it is. Evolution is not the reason of existence; it can describe the means, but it cannot answer why these means exist, or what they exist for. Only a conscious mind can do this. The 'why' only exists for the conscious mind, and it can only be answered by the conscious mind. Biology, as all science, provides facts about life, not what we should do with it. As the great philosopher Karl Popper said, "Nature consists of facts and regularities and is neither moral nor immoral itself ... Facts as such have no meaning; they can gain it only through our decisions and choices in life."

People become so concerned we weren't created as thought they forget the fact they are able to consider this question at all. Evolution may seem a lifeless process from one aspect, but from another it may seem the most mysterious and life-creative process imaginable; what works best with the environment is what evolves, flourishes and is brought into existence. Saying evolution is mechanistic is the same as saying all there is to life is a billion letter linear genetic code, or the only reason we exist is because particular chemical elements like carbon and hydrogen happen to combine together in a particular way. Of course these are true but they're only particular perspectives.

To assert man is the purpose of evolution is more than science can say, but the wonder of all life is here by its principle. When machines can programme themselves, reproduce themselves, source their own energy, and evolve and interact according to their environmental circumstances and influence these circumstances, then they will be akin to the living organism. When machines evolve the ability to sense, remember, think, anticipate, have emotions as the expression of the state of themselves internally and externally, and of what is 'good' or 'bad' for themselves, and form some sense of self, then they will be comparative to animals. And when they evolve to have imagination, foresight, conscious feeling, complex language, a reflexive and extended self-conscious, and a mind capable of introspection and studying itself as well as the outside world, then they will be comparable to humans. One may spend a lifetime becoming an expert on a class of molecules, or the habits of a few animals, but if we consider the whole we will be at a loss. What is worthless in this? It's why humanity is present: a work of hundreds of millions of years, a being of infinite, unimaginable complexity. The product of the precise combination of the laws of this universe; a living, thinking part of life, conscious of its own self and existence. Cosmology may remind us of our insignificance as a small drop in the ocean of spacetime; but our true significance is in value, not in physical location. What more incredible creation could one see than conscious life from unconscious matter?

It is the creation of the conscious mind that has created true meaning in the world, and it is above all upon humans and our choices that this meaning depends: as beings capable of conceiving the reality we exist in and choosing how to live in it. We are born with a human nature and it's the background on which our development occurs, but the kind of person we turn out to be has more to do with our own choices than anything we inherit. Our thoughts and ideas are partly a product of our environment; but whilst the circumstances and context of our lives do much to dictate our thoughts, our thoughts can do much to dictate our circumstances. The real problem for us is not that we're helpless tools of genetics but that we don't use the power we've been given. Every man prizes freedom above all but few pause to consider that if in their behaviour they're slaves to unthinking behaviour they're least free of all. As Confucius said, "by nature near together, by practice, far apart": all people have the same fundamental nature, but what we become depends on what we practice. If we wish to achieve happiness we must think and work for it.

The real virtue of biology isn't that it should lead us to have a contempt for life or human nature. The virtue is that it should lead us to see that whilst we are part of nature and come from it, we are given our conscious minds in order to direct ourselves: to know and direct our behaviour towards achieving conscious happiness. What we should gain from knowledge of the world is not despondency, but amazement, and a sense of opportunity and responsibility. To live for an instant is not futile: it is to be blessed beyond imagination. The point of life is what we make of it, and we can live like the worst of nature and complain we're constrained by it, or being able to think and understand for ourselves we can learn from it and transcend our lesser selves. There is much true joy to be found in life, but it requires effort, self-discipline and right conduct to attain. Problems only arise because we complain of the human condition without disciplining ourselves to surmount it; as the old Sufi saying goes,

"Most of humanity do not know what is in their interest to know. They dislike what would eventually benefit them."

What does science mean?

Science has changed many things. It's not hard to see how people once attributed the wondrous workings of the world to God: the creation of the heavens and earth, plants growing from bare ground, the sun rising from the ends of the earth and stars suspended across the heavens; people suddenly falling ill and being miraculously healed, the earth being split open and lightning striking from the sky. Seeing the sun rise and fall it would be easy to think it revolved around the earth and the earth was the centre of its purpose. But science has shown how these are just the mechanics of physical laws in the universe; how the sun hangs in the sky, why the earth may "give birth to fire," that magnetism and chemistry are not magic, that "portents in the sky" are just orbiting comets, that there is no heaven, just trillions of stars and an infinite space, that plagues are not a curse but just the effect of bacteria that have existed much longer than us.

Yet whilst science might seem to have displaced meaning from the world and rendered many metaphysical beliefs obsolete, there are many reasons why it doesn't answer all the questions man seeks to know, and it's strange that science is seen as replacing or being in conflict with the idea of God, since it's precisely in science and through science that people should seek to know truth about the world. The reality is that science has only uncovered a deeper layer of mystery and wonder to the world. It will never seem sensible to suggest the entire universe and everything in it is just energy manifested in various forms. The next time a man reaches for a cup of tea he will be sure to think the cup is a cup, and the water is real, hot tea, and not a collection of different properties like colour or temperature that belong to a collection of molecules that themselves consist of 'energy'. Or that each time we feel the sun on our face we really feel the effect of energy we call electromagnetic radiation or 'light' that has been emitted by atoms reacting in the sun, travelled many millions of miles in just a few seconds, and is now being absorbed and interacting with the molecules contained in our skin cells, which are transmitting the sensation of 'heat' to our brains via electrical signals in our nerve cells, which themselves consist of thousands of different complex molecules performing different tasks. But all this is what science has shown.

What we call 'heat' or 'light' are just different kinds of perceptions. We happen to see a certain spectrum of 'light' waves, but there are many other waves out there, like radiowaves and microwaves. We consist of a composition of molecules, we walk around in and breathe a gas of molecules, which are themselves particles of atoms and electrons, whose own subparticles are mass or waves of energy whose interaction is governed by the various nuclear, gravitational and electromagnetic forces, which rule everything from stardust to people. We have existed for a tiny fraction of time in the existence of the universe, itself created in a moment, and are no longer even certain how many dimensions there may be in it. Whilst science tells us the most part is filled with an invisible dark matter, that space and time are a single relative fabric, that things are sometimes solid and sometimes waves, and undetermined until we interact with them.

It is science itself that has not only revealed the rich and impossibly deep rational basis of the universe, but shown how improbable it is: from the singularity of its creation to the unlikely set of precise physical parameters, mathematical constants and evolutionary circumstances that have allowed the complexity of nature to arise. This is not even to mention the origin of reproductive life from inanimate matter and the phenomena of consciousness. Perhaps most importantly, science doesn't actually tell us what underlies this reality. We don't know what it ultimately rests on, what 'energy' is in all its forms, or how its laws were created and why they are what they are. These questions must in fact always be beyond science because it is by definition confined to the realm of the empirical: to what we can see and touch and verify by observation and experiment. Hence we have Schopenhauer's comment that:

> No science can be capable of demonstration throughout any more than a building can stand in the air. All its proofs must refer to something perceived [i.e. ultimately unexplained] ... there is a point where every science leaves things as they are ... they always leave unexplained something they presuppose ... the force itself remains an eternal secret ... philosophy alone has the peculiarity of presupposing nothing; everything to it is equally strange and a problem ... Physics is unable to stand on its own feet, but needs a metaphysics on which to support itself, whatever fine airs it may assume. For it explains phenomena by something still more unknown than they, namely by laws of nature resting on forces of nature.

The irony is that we think once something has been described and labelled, we know what it is. But this isn't true at all. The ultimate nature of physical reality doesn't even fit the description of 'matter', let alone our ideas of fixed time or space. It might be said science has shown things like magnetism and electricity aren't magic, but it might also be said they're still the strangest magic a man ever saw, because in fact what really are they? The stuff of this world is a construction of such curiousness that it is beyond all possible imagining, and truly the divinity of life is in science; it is not dispelled by it. What we should gain from science is a little piece of understanding and wonder at the mystery of the world's foundations. Science hasn't diminished the divinity of man; it has vastly expanded the human concept of the divine, and if "beauty consists in a certain clarity and fitting proportion" as Aristotle said, there can be nothing more beautiful than the clarity and interconnection of all the parts of this universe that science has revealed. We should also not forget that science actually has nothing to say about values and purpose in the world or our lives; science gives us facts, but it is left to the conscious mind to decide and work out what has meaning and value.

Chapter 3

What is religion?

Ever since becoming aware of his own mortality in the dawn of human consciousness man has been attempting to find a means of attaining liberation from this state of mortal life. The idea that the end of life is final elicits such despair in the 'I' that when it becomes aware its entire being must be negated, the prospect is enough to cause a great desire to believe in something that will liberate it from this threat, if it's not to go mad instead. The mystery of origins, the fear of what is beyond our power, and the threat of death: that when we die our selves will become extinct, has made belief in the existence of Gods, and in an afterlife or rebirth, common to nearly every society in history. Indeed questions such as these burn so greatly and deeply in man there is always a danger of being convinced of an answer we want to believe. The idea that death brings annihilation of self is such anathema to our nature it's no surprise to find a desire to seek refuge in the ideal: in an eternal life and great purpose to the world. What is less bearable than the thought of permanent separation from those we love and trust? And what greater consolation can there be for life's suffering and injustice than to think that "this is nothing but a passing world; good and evil will be judged in the next"?

But a desire to believe something doesn't make it true. The power of truth is that it is true in reality and for all human experience. As the great philosopher John Locke said: "The unerring mark of the man who really loves truth for truth's sake is the not entertaining of any proposition with greater assurance than the proofs it is built upon will warrant." It's been argued faith is more important than reason, but to reject reason is to undermine any difference between truth and falsehood. It's only by the use of reason that we understand whether we should call something 'true' or 'false' according to whether it agrees with the evidence. Some have feared being rational reduces the value of inspiration and intuition in the world, and limits the creative and free part of man; when it's most rational to value these whilst also knowing where their limits lie. It's only by emotion and imagination that we can empathize and be creative, but relying on emotion without the guidance of reason isn't to attain a greater truth: it is to strike out the criterion of truth from knowledge and confuse imagination with reality. Man and earth are entwined, but this doesn't mean we can decipher the future by looking at stars or fingerprints. The person most wise and free of all is the one who chooses to use their reason above all else to guide themselves on the journey of life, for they are not burdened by arbitrary and useless baggage and guide themselves merely by what is good and useful to themselves and mankind.

However, this doesn't mean we should reject all religion as wishful thinking and foolish superstition. It just shows human knowledge is limited and fallible, that we should be careful what we base our beliefs on, and that our ideas of reality are liable to evolve as our understanding does. There's nothing irrational per se about religious faith: it's reasonable to have faith in things beyond the possible bounds of knowledge. Reason and experience can't tell us the ultimate nature of reality, or whether God exists, or what death brings. But all faith must be informed by rational judgement. Trying to maintain an obsolete metaphysic and refute scientific knowledge instead of embracing it is not wise. It's science: investigation of the world and the work of a rational mind on it, that should furnish a man with his food for thought, not the other way around.

Yet religion hasn't always been oppressive to reason: the pursuit of truth has arguably always been the true purpose of religion. But this truth comes in two guises: truth about reality, and truth about how we should live in it. And unlike the former which must change as our knowledge of the world does, moral knowledge may remain true and unaltered by scientific discoveries. Ethics isn't the same as science: there is no scientific 'truth' or 'falsity' in ethics, only ideas that we agree are better or worse depending on what we seek and how we want to achieve it, since ethics doesn't concern material facts but ways of living. So what was true about people and how they should behave ten thousand years ago may be just as true for us now. Yet in either case, the idea that truth is what we wish it to be and an instrument to be used accordingly is an utterly corrupt idea: deceit and truth aren't compatible. A strong belief, or strong propensity to believe something does not make it true, and any teaching must be examined for what it actually says, since the validity of truth is that it's subjectively true of all human experience, and for all humans: this is how it gains its objectivity.

Religions are a tool which can be used to enlighten or to oppress. Their true use is, as the Sufis have taught, two-fold: firstly to organize man in safe, just and prosperous communities, and secondly, to lead man to inner knowledge and integrity. If our knowledge and understanding change, what of it? Knowledge changes, then so should our understanding and interpretation. Religion's role is to provide a medium for reverence and gratitude of existence, to enshrine moral teaching as the highest good to which humanity can aspire, and a vehicle for pursuit of the truth and seeking some connection with ultimate, transcendent reality: something greater than oneself. But all religious books have been written with a view to their time and place: as temporal metaphor and allegory for the truth. We should keep our minds open and unrestricted by narrow thinking, then we can learn to perceive the truths that apply to our own circumstances, and where the truth that lies beneath all good teachings is to be found. All spiritual methods and teachings are only external means of guiding us to wisdom: what's important isn't the means, but that we should arrive at wisdom nonetheless. Each society has had its own religion and its own means, and each of these has been intertwined with the identity of its culture and people. But the truth doesn't vary: it's the same for all. Therefore, one should adhere to one's own religion, but let it be a means for practicing the essential truth, not a prevention from doing so.

The unbreakable custom

Whilst diversity of culture and religion makes the world an interesting place, this doesn't mean all moral teachings are equally good, and merely because things are the way they are doesn't necessarily mean they are right: tradition isn't truth and custom isn't sacred. Traditions and customs have only been created by other men. Yet what has a long history is not easily changed; time has the power of giving all customs the force of habit. It's often easier to believe what we are told than to investigate things and consider whether they're true or not. But the very value of being human is in questioning and examining all things to determine their truth and understand for oneself; and we shouldn't hesitate to think and know for ourselves and question what we are told. Otherwise in the mass of contradicting opinions and teachings in the world how are the true to be distinguished from the false, and how is one to learn and understand instead of just repeating by rote?

The great disaster in spiritual matters is that many consider the most faithful man to be one who adopts what he is taught most unquestioningly and defends it most relentlessly. But there is no God that will ever punish a man for seeking the truth, and to adopt this nescience; this complete renunciation of questioning and thought as the principle of one's life, is to render oneself as remote from the truth as one could be. The very divinity of humanity depends on its thinking conscious mind, and its ability to investigate, reason and judge without acting blindly like other animals. There is a concept among some that because they follow the ancient teachings they have no need for questions or freedom, but what truth lies in the suppression of knowledge and freedom? Knowledge and freedom *are* the true cause of God. If one cannot be obedient one cannot learn anything, but unthinking obedience is nothing but evil. The mark of true and absolute faith is not following things without thinking about them: it is faithfully using one's mind and reason as they were intended to be used.

As Locke noted,

> Man, considering the various, different, and contradictory persuasions and opinions of men, and the devotion wherewith they are embraced, and the resolution wherewith they are maintained, may perhaps have reason to suspect that either there is no such thing as truth at all, or that mankind has no sufficient means to attain a knowledge of it ... [but] the force of the many contradictory beliefs in the world derives chiefly from the power of what people are taught in childhood, which whether by force of education, or laziness, or ignorance, they do not question, especially when one of their principles is that principles should not be questioned.

Young minds are at their most impressionable, unprejudiced and receptive, and what a child is taught, however unreasonable it may be, will take the force of a principle of knowledge; especially if it seems that this is the most ancient knowledge they hold, and if they observe it to be kept to by all others. It isn't unusual to find people will rather disbelieve the evidence of their own eyes and ears than admit anything disagreeing with such sacred beliefs, whilst they will accept the greatest improbabilities so long as they concur with them. When in fact such ideas are nothing but the first things they were taught by other people. As the Sufis advise, "What you think to be yourself you will become aware is concocted from beliefs put into you by others, and isn't yourself at all ... None may arrive at the truth until they are able to think the Path itself may be wrong." If you cannot think otherwise than you already do, how can you know what you are saying is the truth?

Do miracles exist?

Whilst the 'laws of nature' seem to stay the same, there's no logical reason why they couldn't change or be broken. But the first problem in considering miracles is that if we define a miracle as any event contravening the laws of nature, we must concede there will likely be many laws of nature we don't yet know of or understand, and so it will be difficult to be sure any unusual event is really a miracle. The second problem is evidence. As the great philosopher David Hume taught, all the evidence the laws of nature don't change, observed by billions of people over thousands of years, is so overwhelming that the evidence for any miracle would need to be very convincing. When we also consider the susceptibility of the human mind: the suggestibility of human psychology, the frailty of the senses to distinguish appearance from reality, the ease with which we are deceived by what we wish to believe, and the common desire to witness a miracle or be a prophet from heaven, then it becomes clear how difficult it would be for any 'miracle' report to provide enough proof to be credible. As Hume also well observed, "it strongly presumes against all supernatural and miraculous events that they are observed chiefly to abound among ignorant and barbarous nations or ancestors, but rapidly diminish and disappear in highly developed countries and our present time, to the extent that one must think the whole frame of nature has changed radically from then to now."[4]

People wish to believe in miracles because they take it as a sign of the divine, as if it adds meaning to the world. But the real miracle is already before our eyes, only we don't see it: it's in our own existence and the existence of the world. Habit makes us take for granted what's before our eyes, but what strange force keeps the moon suspended amidst a sky of stars spread out across infinite time and space, or transmits the distant heat of the sun onto the face of the earth, or our feet attracted to the ground? What infinite complexity of biology and chemistry occurs every time we decide to move our arm and it moves, or creates a child from the combination of two cells? The only miracles we really need concern ourselves with are in our existence and nature: the ones we see every day but ignore. As St. Augustine said nearly two thousand years ago, "Although the miracles of the visible world of nature have lost their value for us because we see them continually, still, if we observe them wisely they will be found to be greater miracles than the most extraordinary events."

There is no need to reach for miracles. What we should have faith in is life and its reality and value. Even acknowledging the existence of God does not require evidence of the supernormal: it is in the 'normal' that the true mystery is found. It has been said "mystics exult in mystery and want it to stay mysterious." But this isn't true mysticism: it's superstition. True mysticism lies in the wonder of true knowledge: rational, scientific and moral understanding. There are those who employ enthusiasm and an air of mystery to persuade of the mystical power of beliefs and rituals; but this isn't true faith. True faith is about pursuing truth and learning to live by it in the real world.

We subscribe to superstitions because they appear to express mystery and defy reason. But true mysticism concerns itself with what is beyond reason, not with contradicting it within its realm. It is the illogical nature of superstitions like astrology that reveal their nonsense; aside from a historical record of total inconsistency and inaccuracy, and the fact any child since biblical times could make problematic observations like noting twins conceived or babies born at the same moment do not have identical lives, astrology clearly has as little basis in reason (at least since the advent of modern science) as in evidence. Indeed, as noted by the Sufis astrology is "one of the easiest absurdities to expose, because its followers have allowed themselves to be pinned down to rules." Other superstitions have prudently kept themselves sufficiently vague that they avoid the fate of being rationally examined at all because there is nothing consistent enough to examine.

Chapter 4

Why does evil exist?

If there really is a God, a First Good whose existence is perfect goodness and whose action is perfect wisdom, why does evil exist? God, being perfect, can't have been forced to create the world. And if It created the world out of free choice because it was good to create it, why is it filled with so much pain and suffering; with all the cruelty and violence of nature?[5] It's been argued the world was created so that we may do good, or that good can only exist if its opposite does, and if there was no evil there would be no good, or that evil arises so that good may conquer it; but this just indirectly suggests evil is good because it facilitates goodness, which is clearly not true. There can be nothing good about evil actions, which is why they are called evil, and the world would be better without them.

Yet if we consider God rationally we may see that 'evil' isn't necessarily a problem for the idea of God. What happens to us in life depends on two things: circumstances according to the laws of nature and the actions of men. And the laws of nature just work as they must in order for us to exist: they are consistent and work in a way that has allowed us to evolve, and they do not favour man over any other form of existence in the process, which means earthquakes happen and disease occurs. Real 'evil' only occurs in human actions where there is free and conscious choice to do it or not. We don't call a lion evil because it must kill other animals to survive and we don't call a parasite evil because this is just the nature of its existence. But we rightly call man 'evil' when he willingly harms another without necessary cause.

From the 'perspective' of nature, there is no good or evil; there is no ugliness or beauty. A rotting corpse is the same as a body thriving with life; both are the same cycle of life. In drought

31

men welcome rain, in floods it is cursed. But it isn't the rain that's evil; the rain is just rain. From the perspective of nature, death is the same as birth and birth as death. It's only for men that there are aspects of beauty and ugliness. It's said by some the world must be divine because of its ineffable beauty. It's said by others nature is filled with suffering and selfish exploitation. From the aspect of nature, both of these are wrong viewpoints: beauty and ugliness, happiness and suffering, exist only for the conscious mind. Evil and goodness exist only for the conscious mind. This doesn't mean suffering doesn't exist, but it means in reality it's in the power of men to determine how much evil exists because it is from their actions that true evil derives. It may seem that nature is evil when people suffer and disaster strikes good and bad alike indiscriminately. But this doesn't mean nature is indifferent to humanity: it has created the conscious mind capable of experiencing it. It just means nature allows the created order to exist by following the laws of the universe and it doesn't privilege us over other life in doing so. If we ask why the physics aren't different it's like asking why human freedom can't be limited to prevent harm to others. What we are asking for is a very different kind of existence; one that would be far from what we are now, and certainly not at all like it is, even assuming a different physics and biology were possible at all.[6]

Learning to live with this truth means recognizing that it isn't in our power to control all things, whilst also making use of the intelligence we've been given to prevent suffering as much as we're able. What matters in life are our intentions and actions, irrespective of the situation we find ourselves in, and we shouldn't despair at the state of things but pursue the good and let this be our strength. It may seem difficult to deal with pain and suffering, but in the end a great deal of how much we suffer from pain depends on our own minds. Thus we shouldn't let our spirit become downcast by the difficulties of circumstance: to be downcast is to drag others down; to keep heart is to help others up. Life can be difficult at times, but we should persevere in our pursuit of what is right, of living rightly and honestly, even if it is hard at the time. The right attitude is to persevere whatever the circumstances, and to use these circumstances as an opportunity to practice virtue, never forgetting those more unfortunate than ourselves. "When anything tempts you to feel bitter: think not, "This is a misfortune", but "To bear this worthily is good fortune"," as Marcus Aurelius said. As long as we can keep the light of truth in our heart we may face whatever life may throw at us without fear or complaint.

Why does evil prosper?

It is often complained that evil men prosper whilst the good suffer on earth; that whilst the bad acquire wealth and riches through their wrong, and the corrupt prosper in fame and success, the fate of the good and honourable is to die in obscurity and poverty. But in reality wrong never prospers, and it was not said without reason by Laozi that "It is the way of heaven to show no favouritism. It is for ever on the side of the good man." Aside from the fact that the common fate of the wrongdoer is a life of confusion, trouble and disorder that attains neither success nor prosperity for its efforts, it's delusion to suppose that people who do wrong can ever really succeed. Even amongst the few who appear to prosper by evil one should realize that neither material possessions, nor fame, nor power can make a man happy or give his life value. It is said by the Sufis, "I have never seen a man lost who was on a straight path," and it is just so: if we keep to the right way we will never lose sight of true happiness, and when we dedicate all our actions to the pursuit of truth we will not want for true contentment. But one who does wrong, whether they appear to succeed in life or make a show of happiness, will always be the most unhappy, since one who makes success by lying, thieving or corruption will never be happy where it matters: in their own heart, and this is where real prosperity lies. Without this contentment, one can't attain happiness, and without these, what is life worth?

Tempting one who realizes this truth with corruption is like asking a man who has found his way out of the cave into the daylight if he can be persuaded to return to it; as if he will exchange the sun and the sky for some useless trinkets. Even if you offer all the treasure of the world, it is like offering a heap of dust and stones. So is it said, "Food and clothing can go, but truth must not be lost," for without wealth and power one can still be happy, but without truth, one cannot.

Overcoming evil

It might seem like the best thing a good person could do is to make the aim of their lives to free the world from evil. But it is the absolute idea that some are 'good' and others 'bad' that itself creates hatred. Experiences and the influence of circumstances can determine even the strongest nature, and the problem isn't so much that people are born 'good' or 'evil' but that their experiences and choices in life can make them so. This doesn't mean people aren't responsible for their actions; but it means the more damaged they are the harder it is for them to make decisions for the better and wiser. Everyone has the capacity for harm as well as good; as a Zen master said, "All people have the spirit, it is just a matter of careful guidance. ... It is like water welling up from a spring; block it up and it makes a bog, open a path for it and it becomes a river." Beyond this it is only a matter of power: if people have unrestrained power and lack the wisdom to use it well, then with little effort the harm can be great.

It may seem if violence is a part of men's nature it's inevitable it will have an outlet. But it's not a matter of defying oneself; it is a matter of understanding and knowing oneself. The greatest awareness of existence lies in understanding life not its extinction, and the greatest fulfillment doesn't lie in endangering life but in preserving and fulfilling it. It's easy to indulge destructive instincts and hard to master them, but nobility and freedom are not gained by harming others, and displays of passion are only good when they fight for what is good, not when they move against it. Yet having said this, it's neither profitable nor wise to meet all hostility with love alone and knowing violence makes for unhappiness doesn't mean we can do without it: it is also true that men must sometimes fight to uphold what is right and defend freedom and equality, and in doing so injure and kill other men. We should meet all hatred with love in mind, but if we are forced to confront a problem that cannot be otherwise resolved we shouldn't avoid it; we should do what is necessary to deal with it without hesitation. Yet as Marcus Aurelius advises, "When men are inhuman we should take care not to feel towards them as they do toward other humans." The killing and maiming of men and of human lives is not a glorious thing. It is only ignorance that prevents us from extending our consciousness beyond a fraction to the whole and seeing there is no 'us' and 'them': there is only us.

Chapter 5

The realm of truth

The ancient philosophers believed that men were created with mortal bodies but gifted with divine and immortal souls. Whilst the body and its senses were stuck on earth and confined to scratching an existence, chasing their impermanent desires and occupied with the transient world in which all things change and decay, the mind could get beyond this world to another by the divine light of reason: a world eternally true, free of change and death. And indeed there is part of us which longs for eternal truth and a realm of perfect and undisturbed rest, and dislikes the fact that things in this world die and decay, and we are ruled by changeable emotions and governed by desire. But there is in fact no reason to think that what we perceive with our senses is less real or true than the realm of reason or ideas, since ideas either relate to this world, in which case they depend upon it, or they don't, in which case they are just a fiction that depends on language and logic. And in either case they depend on physical brains to have them; they don't exist in the ether. The long history of philosophers seeking after some 'eternal and pure realm of thought beyond the illusion of the world of the senses' is little different to the history of prophets wishing to believe this is not the real world but it is to follow. The truth is mind is no better than sense: you can't have a mind without anything to experience, and the physical world is the ground of our reality. Our senses and emotion are as necessary as our reason and logic, and we shouldn't think of man as having a "divine mind trapped in a mortal body" since there are as many vices of the mind as of the senses.

Yet this faith in reason led many philosophers to believe that just by exercising their logic and working from innate 'God-given' truths[7] they'd be able to arrive at great secrets about the world: to grand philosophies and mysterious metaphysics. And this is indeed partly true because in logic and maths we can reason our way to truths without needing to study the world; all we need is our own mind. But it's not possible to develop ideas that are true about the world, rather than just true in themselves, without studying the world itself. One can write a book from imagination that may be entirely possible and self-consistent, but it's still fiction in the end: it isn't real because it doesn't relate to reality. If we want to attain knowledge about the world we have to look at it before we start reasoning: we must reason by empirical investigation; by examining and studying the outside world.[8]

As for innate truths these are just a result of being logical by nature. Knowing 'nothing can be and not be' or 'one plus one is two' isn't knowing an innate principle, it's merely a matter of knowing what you're saying and being logical in use of language. Likewise, understanding the concept of good and bad requires an emotional nature, but the concepts themselves only come with experience. We are born with a rational and emotional nature, but knowledge is learned in life and its experiences, not possessed at birth. This is why people's knowledge and beliefs vary with place of birth, upbringing and education, because it's by all these experiences knowledge and ideas are acquired. This also means our understanding is limited in several respects: we can't know or reason our way to knowledge of death, or God, because these are things beyond our experience. Truth is attainable, but only within limits.

The perception of truth and uncertainty

Given this, it might be assumed there isn't much more to know about the world than meets the eye; we can know what we perceive, and what we can't perceive we can't know. Yet the situation is actually even worse than this. All we really know about things is how they appear to our senses, and all we can say is that our minds are given the impression that certain things exist and have certain properties, though whether these things exist independently of our minds and how they have these properties we're not sure. We assume we know what the world is made up of: that this table is made of wood, or that wall is made of stone. We give everything a name and take it for granted we know what things are. But in reality, we know nothing about things beyond the qualities we perceive: wood and stone are just names we give to certain qualities. We don't know what their true nature is or what causes them to have these qualities, and what things are 'in themselves' remains a mystery.[9]

Philosophers once thought of substances as mysterious entities of abstract 'body' in which particular essential 'forms' inhered, giving material things their qualities like colour or taste. Whilst science has shown how physical properties relate to molecular structure, ultimately it's still true all our knowledge of the world rests on sense experience. This means it is hard to be certain that things exist outside us and whether they exist the way we assume they do. It's only because a reality we can't control appears to exist (we can't conjure sensations or choose which ones to have) and because we appear to share it with other people that we assume an 'external reality' actually exists. But what we really know about this reality is limited and has little to do with common sense: it isn't composed of solid matter, or even eternal atoms or particles, but energy, or the manifestation of energy in particular events. There is no fixed or absolute space or time; these are relative and only the framework in which events are manifested to our perception. And the point at which we can distinguish the world 'in itself' from our perception of it is not easy to find.

As Hume also taught, there's also no logical reason to assume that what we've experienced of things in the past will be true in the future; there's no logical reason why the laws of the universe must be one way rather than any other, or why they shouldn't change in future. We have no idea what logical framework underpins the existence of reality. Experience tells us it's almost certain the earth orbits the sun due to something we call 'gravity' and in the morning we will see the sun rise over the horizon. But our only proof for this is our past experiences and our induced knowledge of the laws of the universe. And not only are the 'laws' of science just our best guess of how things work, no matter how many times we observe them to happen our resulting predictions will still only be probable, not certain.[10] Indeed, Hume questioned even this, arguing that we have no logical reason to assume that the laws could not change and the past is *any* guide to the future. Modern science has provided a means to uncover layers of reality that we can't directly perceive and dramatically increased our knowledge of cause and effect,[11] showing that the laws of the physical universe operate as an integrated fabric in the process. But all scientific hypotheses are still based upon experimental observation and depend on experience, and this means we can never be certain any 'law of science' or experience is a certain guide to the future.

The same largely holds true for any extreme skepticism or 'relativism': the philosophical view there is no absolute truth at all. Because so little is certain and there is ultimately no other guide to the truth than the mind (or collective minds if we are inclined to believe others exist), it's not possible to easily refute such skepticism, if at all. There are degrees of uncertainty in life; little in logic, more in science and perception, even more in ethics; but absolute certainty is in short supply. Because of this life requires a leap of faith in the 'truth' of our reason and experience if we are to avoid falling into doubt and confusion about everything. We must have faith that our senses don't deceive us, that the world and other people exist, that the sense impressions we have represent something real, and that reason works: that we are able to judge true from false and interpret something resembling what is 'real' from them. And we must have some faith in the strangely consistent laws of nature, in inferring probabilities from induction and in using the past as a guide to the future.

However, having this faith is not at all optional: anyone who chooses to live must choose to live by it. Most students of philosophy would perhaps agree with Hume that philosophy is a one way voyage that takes you from the comfortably certain seaside village of common sense into the broad and deep ocean of uncertainty and doubt, from which only this faith can rescue you. Yet this is not a reason to become in any way depressed, since as Hume also recognized, everyday reality keeps working just fine even though we know very little about it. Moreover, there is nothing to prevent us trying to expand that little in the meantime, and the success of science in not only predicting forces of nature, but harnessing them in aeroplanes and medicines suggests that is not a futile endeavour.

What is the point of philosophy?

Philosophy can't gain knowledge of death, or total reality, or 'things in themselves', or God, and it can't ultimately answer questions over the meaning of life. So what is the point of philosophy? Firstly, to realize that things aren't just what they appear to be and to make us think about their nature. Fundamental problems of knowledge exist, reality is not what it appears, proving the external world exists is difficult, the process of perception-experience is strange, ultimate reality appears to consist of 'things-in-themselves' that have little relation to common-sense or how we perceive them, the origins of ethics are hard to untangle, almost everything is beyond certainty, and our knowledge of the universe has many paradoxes. Why does anything exist and how is regularity and rationality built into the framework of the universe? What is consciousness and the relationship of self and other? What is ethics? Philosophy has few ultimate answers, but it helps us to understand the problem of what is and what we are, and isn't totally without solutions.

Secondly, philosophy can make us wary of what habit and custom unconsciously impose upon us and experience makes us take for granted. 'Common sense' causes us to intuitively make broad assumptions about how the world works and what we should do in it. But if we want to consider what the *real* nature of reality might be and what we should *really* be doing in it, we need to stop thinking uncritically and start thinking philosophically. Habit makes us complacent about the world unless we make an effort to pause and consider it; but it can be useful occasionally to shed the temporal illusion of phenomenal reality, and the idea things around us we touch and see are permanent and stable, since they are not at all. In this context, the philosophy of ethics is perhaps its most important use of all: considering what really is of importance in life and how we should live it. We assume we should pursue what is good for us; but deciding what is and how to do so are not straightforward questions.

Thirdly, understanding the importance of assuming nothing, questioning everything and the 'critical rational' approach in developing knowledge: having a reasoned and critical basis to belief and understanding. Rationally considering the evidence for and against what you are arguing. What is the evidence we are right and how could we be wrong? How can we achieve clarity and consistency in knowledge whilst also understanding its limitations and uncertainty? Science has provided the best and most tangible evidence we have of a deeply layered and complex reality that exists beyond the realm of direct perception. But if you don't understand some philosophy, it is difficult to understand what science is.

Fourthly, humility. Learning about the limits of our knowledge teaches us we should be both free of complacency and arrogance and wary not to take the world for granted but pause to consider the marvel of its true nature. It also teaches us we should use skepticism as a guide to believe things in proportion to the evidence: we should be free of dogmatism, especially in matters beyond direct experience, keeping an open, receptive and unprejudiced mind. Philosophy shows that reason itself can't do without some faith, but faith should always be determined by reasonable thinking. We should appreciate what existence has given us, think for ourselves, strive to know what we can, and recognize the mystery of what we cannot: neither ignoring it nor assuming we can know for certain what is beyond our power to discern.

Chapter 6

Free will: the insoluble paradox?

If we're free to do as we please what place is there for God; for an omniscient being that controls everything? But if our fates are predestined how can we be responsible for our own actions; since if our futures are already decided what use is it for us to try and act other than we're already destined to act? If there's such a thing as God: all-encompassing and all-powerful, whose knowledge covers the history of all time, then surely everything must be governed by fate and happen as it will? What futility is there in us trying to change the course of what is already ordained, and what room for our wills to breathe? But if God doesn't exist, and we are really free to act, then we are responsible for our actions; but what does it matter what we do since there is no God to judge us and no absolute law?

The first answer is that neither is true because both make their ideas from a human conception of God. If all men were predestined to their fates, to do good or evil, why would some be created only to be condemned, and what is the use of talking of morality when we can only act as we must? The idea of morality depends upon us being free to choose our own actions and being responsible for them. Yet it's equally clear we aren't completely free to act as we please, and our power is limited, and many things which affect us are beyond our control.

The right answer from man's perspective is that we are neither wholly free nor wholly powerless. We're constrained in our genetics and environment. We can't choose to feel emotion or be born with a human nature. We don't choose to whom we are born or in what circumstances. Yet every person is born with a conscious mind, and is free to think for themselves and as much as they wish determine their own behaviour. We may not choose where we find ourselves but we can choose how it will turn out: we can choose to imitate the conduct of the wise and virtuous, or we can choose to imitate the behaviour of the ignoble and corrupt, and this depends on ourselves alone.

The right answer from God's perspective is that there is no change, or time, or process for God. If God sees or directs all that is and will ever be then it must be in an eternity in which there is no time: no present, nothing to come before or to follow after. But this knowledge, if it exists as 'knowledge', isn't ours: we are free and every decision we make is our own responsibility, for the future for us is unwritten and the time hasn't passed yet. The knowledge of what will come is not ours to know. We exist within space and time, and our destiny can only be fulfilled by our acting and choosing as if we are completely free; as if our fate is in our own hands. It's a mistake to think there is nothing we can do: "Whether I shall be saved or lost is already decided, and so this cannot be changed whatever my actions." It *is* our actions and efforts that determine our fates.

As the philosopher Leibniz taught, God's mind may be infinite and have an understanding of each thing so complete it can instantaneously perceive everything that can be attributed to it, and everything that has happened and will ever happen to it, every cause and effect associated with it. But we don't have the power to determine the truth and nature of things from mind alone; we can't see things and instantly deduce their every property, or what their past, present and future hold. We don't have the privilege of seeing from a state without beginning, continuation or end. It may be that we are part of the chain of causes, but the chain only leads up to us, not beyond. A man's fate isn't inevitable or unavoidable; many fates are open to us, and that which finds us will depend upon the strength with which we seek it.

It may be argued that even without God we have a dilemma. For doesn't the universe operate by physical laws, by cause and effect; don't all effects have causes, and aren't all human actions the direct consequence of circumstance, motive and character? Aren't all part of the infinite chain of causation that inexorably determines what happens?[12] But this is just a word game. Every effect may have a cause; but in the sphere of human activities no effect can be predicted from any cause without uncertainty, and no effect is predetermined with certainty from any cause. It is still the conscious mind that must choose which effect it wishes to seek as the outcome from any cause, and the freer the mind: the more it is freed from its own instinctive and conditioned responses, the freer its choice over which 'effect' is the result. No conscious action is 'controlled' by preceding events: circumstances may predispose us to choose an appropriate course of action, as they should. But the only event controlling a human action is the one that occurs in the mind when it deliberates over what to do. It should be obvious to every sane individual that they are not totally free in life: certain things motivate behaviour, others don't. And that behaviour is limited: we can move our arms and legs, we can't fly through the air or bend the rules of physics. It should be equally obvious that partial freedom is not the same as no freedom at all.

The truth is that we are as free as we wish to be; and it is through every choice we make that we continue to determine ourselves for better or worse. There is no law of history; no fate that ensures good will be victorious or the inevitable course of the world. History is as full of evil as good; as full of tyranny as justice, because destiny depends only on ourselves and what we make of it. Nothing happens in life without purpose, unless we should choose it to be purposeless. The proper attitude for the student of wisdom is to "live with cause and leave the results to the great universe." This means we must do what we can, and whatever is in our power to do, whilst always knowing it is beyond our means to determine the outcome of all things. What will be is the will of heaven, or what you will call it. What matters is that we do what we can. As the Sufis say: "What is done for you, allow it to be done. What you must do yourself: make sure you do it."

God

The world's variety of metaphysical beliefs about the nature of God and ultimate transcendent reality cover the entire spectrum of possible alternatives.[13] In the West ancient beliefs about the immortal soul and mortal body, and the conflict between the rational mind and irrational desires led the philosophers to consider that soul or mind represented the divine in man, whilst the body and its desires represented what was mortal and earthly. So God was considered a supreme and perfect form of the rational principle of mind or soul. The First Being; underived and uncreated, and the source and cause of all things existing and having goodness. Plato and Socrates had their supreme 'Form of the Good', in which all the other forms of the world ultimately share and by which they have the quality of being good. Aristotle called God the "First and Unmoved Mover," responsible for initiating all change and process in the universe whilst standing outside of it and remaining still in the perfect activity of "contemplating contemplation." Successors saw God as the perfection of all attributes: perfect in His Being, Truth, and Unity, and the essence of all these forms of perfection and the cause of them existing. Eternally good; the perfect rational entity in a perfect state of immortal happiness, and the supreme reality in which all other things subsist according to His one essence. The Idea of all things and the Author of all Forms, not only present to the whole of time but containing all time: timeless, changeless, a perfect spiritual unity. Others have seen all reality as a manifestation of this divine will, or more negatively as an illusion masking it, or dispensed with the theory of a person-like-nature for transcendent reality altogether.

But in contrast to the varied and sophisticated abstraction of religious philosophy, the popular forms of religion have often been something completely different. A fusion of myth, superstition and culture that has often arrived at Gods little different from the humans supposed to worship them, much to the confusion of what is meant by God; and to the confusion of many that they believe they can say anything about God's nature at all. As every religious teacher from Bodhidharma to Aquinas has known, there is nothing at all we can say about God's nature. Yet this fusion is no surprise given it's easier to imagine God in human-form, possessing all the attributes and powers desirable to us, rather than trying to conceive of some infinitely perfect entity to whom all time is instantaneously present, and to whom every thought and action is instantly perceived and understood.[14] Hence it has been the aim of every religion to introduce the concept of the divine by the familiar and conceivable, whilst reserving the true concept, that there should be no concept, for its esoteric core.

Yet even if God is talked about in symbolic terms how could the cause of God involve hatred or oppression, or being deceitful or false? How could a being of infinite perfection support injustice or inequality, or 'desire' to be worshipped, or the God of a particular people? How could the salvation of man depend on being religious for fear of punishment or the promise of reward? Why would God; the omnipotent, omniscient, free and infinite creator of the immense and complex universe, choose to 'reveal Himself' especially to a small group in a particular part of the earth at a particular point in history, or grant special powers to any particular intermediary? Would not God rather ensure that every conscious living thing, whatever point of time and space it happened to exist, could receive this perpetual revelation simply by looking inside themselves and at the world around them? Would this not be a better means to ensure total equality and value of existence for every conscious creature that comes into being, whether that happens to be an Australian aborigine of sixty thousand years ago or a twenty-first century New Yorker? That, as Kant said, "the starry sky above us and the moral law within us" are clear signposts to the truth for anyone who wishes to observe them?

Men have often conceived of God in the mould of themselves, but God couldn't have emotion, or intelligence, or being, or existence, or perception; or any such thing in any possible way we can use these words to mean. Any attempt to describe such things in God's nature is meaningless. The finite can't describe the infinite. God is not a being; God is beyond time, beyond space, beyond action. If we are still thinking of being or not being, then we are not thinking of God but thinking of men and the world. There is no before or after for God, no time to pass or future to come; past and future just exist for men and the world. Being infinite and beyond time and space how could any such concept apply? God isn't some omnipotent being in the mould of the human mind, or a conscious designer of the universe but something far more subtle and infinite. And the real harm of ideas of 'God' is when instead of inspiring men to virtue and wisdom they are twisted into a cause of the very things they should be used to overcome: violence, superstition, and dogmatic intolerance. When people destructively project their own fears and desires onto a constructed and imagined deity, instead of realizing the transcendent nature of reality that should lead them to overcome their own limitations.

Spinoza said that God is the single absolutely infinite substance of which the universe consists, and that all things in the world are finite modifications of Its infinite attributes. This doesn't mean It is the foundation of some shared perceptual reality, or the energy that forms the sun and stars, since It is itself neither material nor mental; but It is the nature of both. This shouldn't sound impersonal: there is nothing preventing us from being as connected with that of which we are a part as we wish. "The human mind is a small finite part of the infinite attribute of the intellect of God, and the human body a small finite part of the infinite attribute of the matter of God." As are all things, from the living trees to every photon of light emitted by the sun that causes them to exist. As a philosophy this has flaws and lacks proof: but what use is theory in the realm of poetry? As it is said in the Kena Upanishad, "What cannot be spoken with words, but that whereby words are spoken; know that alone to be God." If we wish to try and understand God we should try to fit our mind to the universe rather than fit the universe to our mind; "You live, move and have your being in It. It is within you and without you, and you are made of It, and you will end in It. It surrounds you as your own being."

There is no reason why the universe should exist, but it does. There is no reason it should be governed by a logical structure of symmetry and beauty, or why evolution should have arisen and culminated in the self-conscious mind. The most that might be logically said is that God is the reason why something rather than nothing exists, since our existence need not be, but it is. Yet it isn't necessary to prove God exists,[15] and there is no evidence that can do so. The facts only lead to an abyss beyond the visible; a great unknown beyond the realm of the known into which only a leap of rational faith can carry us. All that is necessary is to acknowledge there's a great purpose and value to life if we choose to realize it, and the only thing we need have faith in is ourselves, and in the purpose of living and its use; which is to realize that the reason we are made as independent beings is to take responsibility for our own lives and to live them as we should: to pursue happiness and fulfillment for ourselves and those around us. Realization of God, whether you choose to call It 'God' or some other name, will then come naturally enough of itself.

Death

Even if we acknowledge a purpose to life we still face death. In the span of all time our life is an instantaneous passing and each minute draws away its short thread. What hope can there be when death will cut it off? When 'I' will be destroyed? You can't separate the self from the brain; from the head on our shoulders. It evolves with the body, depends on the body, and is inseparable from the body. So it seems there can be no mind without body and it's futile to hope for an existence beyond death: it must be the extinction of life and nothingness. Our wealth and works we can't take with us, our body will be food for the earth, and we must be parted from all we love and trust. Should we have faith in eternal life as a delusion: for the sake of our happiness and sanity because the alternative is too futile to bear?

But the answer is neither to hope nor fear anything. The wise attitude is to influence what we can and reconcile ourselves to what we cannot. Our birth wasn't in our power and neither will be our death, what then does it matter? We don't fear the non-existence that came before birth. We don't fear the sun rising, or that we must breathe. We shouldn't then fear death or complain that others must die. Besides, we are so concerned to avoid death without considering what it means, for if we could survive another thousand years, or ten thousand what difference would it make? What meaning would it add? To think of the purpose of life as some future thing to be achieved is a mistake: the future can bring no such thing. There must be a purpose for us in this moment or none at all. Even creating the next generation is not purpose. The line of human beings will end one day; humanity will end, time will end. The meaning of life for us is here and now.

Realizing the profound worth of each moment can lead to a dangerous anxiety. How can we deal with knowing each moment will not come again? But the answer is very simple: so little is in our power and so little certain that we shouldn't be concerned. We can always take the limited human condition as an excuse for failure: dissipating life in frustration and futility, and blaming the world and everyone but ourselves for our situation. But life is a gift; a precious and rare opportunity. The point is not to be angry because we are going to die or because others must. It's to appreciate and be aware of what it is to be alive, and to use the life we have as best we're able: not wasting a moment. We are all concerned as to whether we'll live or die, and what will happen when we die. But all must depart when their time is due, and our only concern should be this accords with the natural way of things; parent before child.

We actually don't die at all. Our small selves are temporary but the force that constitutes our "true indestructible inner nature"[16] and the world lives on. Existing as a short manifestation of the universal that will shortly welcome us home, we should be utterly free from fear in life; not reckless but fearless. It's been said, "the way to enlightenment is through the ordinary mind, and when we are enlightened, every moment we regard as prosaic and dull in our deluded condition becomes wonderful and divine." If we wish to seek this enlightenment for ourselves; to cultivate an unshakeable joy in existence and a faith and perseverance that are undisturbed by whatever we may encounter in it, even death, we should pursue it with all our heart and mind.

Part 2: Resolution

All the grand sources of human suffering are in a great degree, many of them almost entirely, conquerable by human care and effort.

J.S.Mill, Utilitarianism

Chapter 7

Resolving to pursue enlightenment

Even though most of us realize that wisdom and virtue aren't easy to achieve, we also have an idea we are born with all we need and expect happiness to come freely. But nothing is free in life and we aren't born as complete and wise human beings: we have to work hard to achieve self-improvement and happiness. Many make grand resolutions without getting far, but we shouldn't make resolutions and expect anything to change by itself; if nothing of value is attained without effort and discipline how much less can one expect to attain those things that are most valuable unless one has the will and discipline to be worthy of attaining them? It must be as if "one has gulped down a red-hot ball of iron which one can neither spit out nor swallow," as it is said in Zen. It doesn't matter if your ability is less or circumstances difficult. Difficult circumstances make for good learning, and if you're led to practice with single-minded resolve they'll be a help, not a hindrance. As the proverb says, "Poverty is your treasure. Never exchange it for an easy life."

Like resting on a giant stack of brushwood when the bottom has been lit, we often pay the question of existence no attention, thinking we'll live forever. But life is finite and we have a limited time to make use of it. The world is filled with enough mediocrity and falsity: we should make it our resolution to eliminate faults, to pursue what is good, and to make ourselves an example for others to do the same. And we should live a full life but know what it consists in: making our living in the world, enjoying freedom without neglecting responsibilities, being mindful in conduct, and learning to be in control of ourselves in pursuit of what is good: in desire and in action.

As for mistakes, the right attitude is not to despair in them but to learn from them. People continue to do wrong because they think that once they've started there's no return, and resolving to change is beyond them. But wisdom lies in doing what you can now, not a futile regret of what's past, and we can resolve to change any time we wish. If we've committed an error we should resolve to repair the harm and never again to do anything we'll regret; and this choice of a way of life is open to anyone at any time. Yet it isn't enough just to regret, since this usually leads to depression and inaction, where one can see no way out of one's situation and no way to change. One must regret and reach up, vowing to pursue honesty and turn away from what is false. With self-examination one should conduct a sincere and profound reflection on how one can pursue virtue henceforth, and then do so.

The delusion of desire?

All things are pursued to satisfy desire, but the indulgence of most desires doesn't satisfy the appetite: the more we have them the more we want them. Or they turn out to be illusory: when we achieve them we find they're not what we wanted at all. Thus the first step to achieving enlightenment is to achieve some control of oneself: ordering one's body and passions to the control of one's mind, and the mind to the pursuit of one's better self. As it is said in the Upanishads, the body is likened to a chariot, the self to the charioteer, the mind the reins, and the horses to senses, and their paths the objects of sense.

We shouldn't mistake what is meant by this. The senses are the source of most joy in life: they are how we hear, see, touch, smell, taste and engage with the world. But a rational use of them is necessary. Sensual pleasure is in the right way essential to life, but without some order over our body and senses, as well as our mind, we won't achieve happiness. Indeed, the great cause of evil in the world is man's lack of self-control over two desires: desire for sensual pleasure, and desire for power. The former is the cause of all kinds of greed; the latter is the cause of all pride and egotistical delusion. Like all animals we react emotionally to pleasure and pain. We can't change this, but we can choose how we act from it and how emotions develop. All higher animals have emotions of fear, desire, enjoyment, sympathy, embarrassment and pride; but only in humans is the mind so developed there are purely mental 'feelings', and not only these but awareness of having these and the ability to reflect and control them. In man not only do emotions cause mental feelings, but mental feelings can cause emotions; we can become so happy or depressed over some idea that it causes the emotion of pleasure or pain. So the ideas we develop and pursue can determine our desires. And so all mental disturbances; all desires, anxieties, guilt, anger, insecurities and fears can have a powerful and direct effect on physical and mental well-being.

There is no question of ethics, or conscious happiness, for other animals. When they follow their instincts to survive, live and reproduce, they obtain their maximal well-being. But in humans there are many desires and the greatest desire of all, happiness, isn't easy to attain, for different desires often contradict each other and chasing them like any other animal doesn't yield contentment but destroys it. Animals are compelled by their appetites, and their only peace is to follow them; but the peace of man is only attained when he thinks and acts in accordance with what he knows is right. The result is that there is nothing humans seek so much as true happiness and nothing that is so hard to achieve. As St. John of the Cross expressed,

Miserable is the fortune of our life, which is lived in such great peril and wherein it is so difficult to find the truth! For that which is most clear and true is to us most dark and doubtful; wherefore, though it is the thing that is most needful for us, we flee from it. And that which gives the greatest light and satisfaction to our eyes we embrace and pursue, though it be the worst thing for us, and make us fall at every step.

Humanity has the potential to be transcendent of the nature that has produced it, and by possessing self-consciousness is already separated from what it has evolved. But it's only when men act with control and understanding in their actions they can aim at happiness and enlightenment. And the key to achieving this is learning to control and govern desire by making all desires secondary to one governing desire: the desire to attain truth and live by it. As Rumi, the great Sufi, said; "the degree of necessity determines the development of faculties in man … therefore, increase your necessity." If we wish to attain perception of greater truth we must only make it our necessity to do so, and to be free of those things that prevent us from doing so; then we will.

Dedication

The two rules for making any serious decision are that we should give it serious consideration, and having decided, we should stick to it. In resolving to attain enlightenment this means firstly reflecting deep in the soul on what it is that we desire to achieve in life; secondly on what is necessary to do this, including an examination of ourselves and what must be changed to achieve it; and thirdly, resolving to pursue it until we achieve it or death prevents us.

We often fail to distinguish between appearance and reality, but nothing of true value is achieved without effort and there are no easy paths to success or happiness. If we wish to grasp the supramundane we must have the will to persevere when the way seems lost and when the temptation is to leave off our pursuance. We must keep firmly with this resolution all through the path of life and know it doesn't happen at once but over much time with great effort. As an eleventh century Zen master taught, "The will should be made single minded, unregressing, for a long time. Then someday you will surely know the goal of ineffable enlightenment. If, on the other hand, the mind retains likes and dislikes and you indulge in prejudice, then even if you have a determined spirit like that of the ancients I fear you will never see the way."

We should learn to order our disordered inclinations and be watchful for obstacles to this resolution; dejection, laziness, self-pity, procrastination and lack of concentration. And we should pursue virtue with perseverance, patience and energy. It is the series of events, of thoughts and actions that make us. Whatever these constitute, they are all we are; then we should not forget it's in our power to strive to make these, and their series which is our life, driven to a good purpose. Only one who has this faith can learn to dedicate themselves to the cause of what is good, and to find joy in all work so long as it's toward this end. Without the illumination of this faith life is lit but dimly by the restless desires and disquiet of an unstilled mind. But when there is faith in reason and virtue, we may find what we seek.

Part 3: Pursuance

All genuine virtue proceeds from the immediate and intuitive knowledge of the metaphysical identity of all beings. However much my individual existence, like that of the sun, outshines everything for me, at bottom it appears only as an obstacle which stands between me and the knowledge of the true extent of my being.

Schopenhauer, World as Will and Representation

Chapter 8

The law of morality

The view that morality is something we ought to do rather than something we should want to do has led many to consider that moral worth can only come from 'duty' and involve suffering and self-sacrifice. But the only real measure of moral worth in a person should be a conscious desire to do good.[17] A moral action needing self-sacrifice may be even better than one that doesn't, but being moral doesn't mean always doing things that involve self-sacrifice or difficulty. It just means being sure to choose those things which are best: doing whatever increases the happiness of others the most. Whatever will create the most well-being and relieve the most suffering.[18] The rule shouldn't be just 'do as you would have others do unto you'. It should be do as you think is most perfect; whatever would be the most virtuous and admirable way, do it that way. In resolution and in pursuance, aim for the ideal that has no regrets. And the best man isn't one who is greedy, proud or jealous, but moral despite this from bitter duty or promise of reward. It's the man who is wise enough to free himself from greed, pride and jealousy and perceive that he should be moral because it is right to be so. Likewise, the true use of morality isn't in keeping up the appearance of good behaviour and we shouldn't be concerned to appear moral whilst underneath seeking whatever we can get: the true use of morality is that happiness and enlightenment depend upon it.

Because conscience seems so powerful and in conflict with the rest of man some have thought of it as the external guide we may follow when beset by difficulty, temptation and uncertainty. But what this conflict shows is that we are subject to many conflicting desires and there's a need to achieve self-control and follow those desires which are really good for us. If we didn't have many conflicting desires the idea of ethics wouldn't even arise; if our desires led one way and the 'law of morality' led another, there would be no ethics because everyone would simply follow their desires. The idea of ethics only exists because we have many desires, and some lead to morality and happiness, and some do not.

Some ancient philosophers believed moral concepts were innately planted in our minds by God, and by being led by our reason to obey them in spite of our desires we were led to be moral. So living rightly meant following 'God's laws' instead of our own desire; rejecting what we wish to do and choosing instead to follow the external law of God as the means to salvation. But whilst this is partly true and a very good guide to personal conduct to one who understands why it is true; it doesn't fully explain why a man should obey such a principle, or why he should feel that he ought to follow it, since a man may just as well decide he should only do what he wants, and one will have difficulty in logically arguing otherwise.

The reality is that morality depends on both our reason and emotions, and it is something within us and our nature not external to it.[19] It is because it's written into our nature that if we wish to be happy we are obliged to follow it. One can't dispense with all desire and inclination in morality and replace it with pure reason; one can't speak of ignoring what one desires and following some external law. It is only by a 'desire' that we do anything at all, and the fundamental basis of morality is emotion as much as reason: the heart as much as the mind.[20] Morals follow from principles of feeling innate to our nature and the reason we think of something being right or wrong is because of both the way we think and feel about it. Logic provides the framework for how to act, but emotions and the desire for happiness underpin it.

The true 'law of morality' is that which has been written into our being: in our emotions, reason, intellect, memory and perception, and if these are used properly we will be moral, and we will be happy. This is the only law that God has written for us. It has been taught that in order to be moral it's necessary for us to reject all desires and do instead what God commands us. But morality only depends on following the truth within ourselves; on using our reason and emotions to judge properly and pursue the true fulfillment and happiness that is the purpose of every human life. It isn't necessary to reject all desires; it's only necessary to control them, to distinguish amongst them, and to reject those that are detrimental. And this decision of which to follow, and whether to be moral, is only our own and rests only on our own decisions. As Popper said, "the responsibility for our ethical decisions is entirely ours and cannot be shifted to anyone else; neither to God, nor to nature, nor to society, nor to history."

Human nature is both self-interested and benevolent; but whilst benevolence may often seem to conflict with self-interest this is an illusion. If benevolence is obliterated happiness isn't possible, and the true self-interest of man is to take an interest in others just as much as himself. A man should learn to think of others as himself, and of himself as other people. People who are purely self-interested may appear to love themselves most, but do no such thing, and must in reality dislike or even hate themselves because they cannot find happiness themselves and they make no one else happy either.

What is right and wrong?

The old debate of intuition versus experience in morality; whether it's something we 'just know' or something we learn, is not insoluble. We are not born with complex ideas: toddlers do not pronounce on the complexities of moral law, nor do they often follow it. But we are born with a rational and emotional nature that inclines us to pursue what is good for us and to avoid pain, to feel emotionally and judge rationally that others have the same right, and to reason our way to judgements about how we should do this; including which pleasures we should pursue and which are illusory, harmful, or counterproductive. We thus have the intuitive capacity to be moral beings, even though the material on which we hone our moral wisdom only comes with life-experience.

It is because morality depends on both ideas and emotions that people from different places and backgrounds often disagree over what they think to be right or wrong, and change their views in line with their knowledge and understanding. However it's also true that all people share the same fundamental nature and find the same fundamental things good or bad, because these things make people happy or unhappy all over the world. Yet since they aren't 'facts', it's true that ethical ideas aren't 'true' or 'false' in themselves: they are just opinions on how to live. So what is ultimately right and wrong; how can we avoid people saying the law of morality is whatever they believe it is or whatever they feel like?

We can decide by seeing that moral ideas are right or wrong depending on how they agree with human nature and what we are aiming at. And since all people are made more or less the same, and since happiness is what everyone seeks, 'true' morality means doing whatever promotes true happiness. Everyone's ethical judgement is subjective; but a man can make his moral judgement as objective as possible if he assumes other people exist, they're made the same as himself, and that the same things make them happy as himself. If he reasons that all others are born with as much of a right to be happy as himself; that all others are as much "ends in themselves" as his own self.[21] And the way to avoid practical solipsism; of being left with our own arbitrary and subjective moral judgements to guide us, is by communicating with the outside world in order to check them: to ensure that they are true because they do really help to reduce suffering and improve well-being in the real world. History is full of grand ideas that seemed 'good' in theory but created only suffering for people in practice.

The divergence of moral opinions and conduct in the world, and their chequered history, shows that merely claiming moral principles are self-evident or intuitive is of little use. There must be a way to judge between moral opinions that doesn't just rely on opinion. Even if a divine law exists we must have some way of judging whether we've understood and interpreted it correctly. And the only way to do this is by reference to the effects of those actions; by whether they produce pleasure or pain; happiness or sadness. This is the guide that God has written for us: what makes us truly happy is good; what causes us real pain is bad. Some dislike the idea that morality should depend on seeking pleasure and avoiding pain: that life's most noble aim is to increase pleasure and minimize pain, since this seems to undermine and debase the idea of morality. But this is only true if our idea is a blind and selfish pleasure rather than real pleasure, which is not really 'pleasure' at all but happiness of the soul. The fact is that it is pleasure and pain, or more properly, happiness and sadness that not only do already motivate all human action, but that should be the guide to all human action. There is no morality that can function without making some ultimate reference to the consequences of actions: to the effect they have on this quest for human happiness. It is because of this that any attempt to formulate a purely rational guide to morality will fail, because reason alone cannot distinguish good from bad: morality also requires some reference to emotion; to happiness and unhappiness. It is also because of this that the real question of morality is actually not at all whether we should judge morality by happiness, but what we judge real happiness to be and how we can achieve it without being derailed by weakness, error or illusion. The tricky part is not realizing that happiness is the purpose of life and morality: it is realizing that sometimes good things require pain to achieve, and sometimes things that seem good only cause pain in the end.

The conditions for happiness

The result is that aside from some common human emotion, morality just depends on two things: firstly agreeing that happiness is better than suffering, and secondly that everyone is equal. No one has any trouble agreeing with the former. And the second just depends on seeing that all people are born of equal worth and there's no rational basis for hierarchies of caste or heredity, and no reason why one person should be considered to be born superior to another. Accepting justice means accepting equality before the law for all. The only problem then left to work out is: what social conditions will create the most happiness? And the only answer, as testified by experience and supported by reason, is that the only condition for a sustainably prosperous society is a free and just democracy that aims at continual self-improvement. And the simple answer to why this is the case is because a representative democracy is the only system that aims to universally protect individual rights and freedoms rather than selectively exploit them, whilst allowing regulated but relatively free market forces to create an efficient and prosperous society rather than an inefficient, bureaucratic patriarchy. It is also the only system that reins in the institutional imperative to grow in size and authority, which in government as elsewhere, rarely has a healthy effect.

Government and institutions are required to organize society but in reality there are no "divine and wise kings" to rule the people, just other people, and thinking otherwise is a dangerous illusion. It has long been the dream of men to idolize and deify the infallible and divinely appointed leader, but no mortal is infallible. If power is granted without restraint or subjection to democratic law,[22] the happiness of the majority will not result except in exceptional circumstances. One may argue about the merits of leaning 'left' or 'right', but the lesson of history, as much as philosophy, is that if we wish to be morally 'true' and objective, we must be democratic. Historical movements against democracy have mostly arisen because either they've never really agreed with the idea of human equality, or because they haven't thought practically about a system that will facilitate universal prosperity. But as Popper well summed up, "Romanticism ... may seek its heavenly city in the past or the future; it may preach 'back to nature' or 'forward to a world of love and beauty'; but its appeal is always to our emotions rather than reason. Even with the best intentions of making a heaven on earth it only succeeds in making a hell; that hell which man alone prepares for his fellow men."[23]

It is no surprise to find in every period of history the writings of men lamenting the current age of ignorance and corruption and harkening back to some former period of glory and wisdom, when the rulers of men were wise and good, the times were plentiful and prosperous, and there was no evil or ill to be found. But in reality the tale of human history has mostly been one of oppression of people, of much unrestrained greed for power and wealth, of sickness, tyranny, ignorance and superstition. And its passage has mostly been one of improvement in standards of living, degree of civilization, in education and literacy, and medical science, in individual rights and welfare. There has never been a time when men were "braver, better, and simpler, unspoiled by wealth, and undisturbed by ideas," and socio-political sophistication in a democracy is the only means by which freedom, equality, prosperity and justice for all may hope to be achieved.[24]

Chapter 9

The wisdom of Spinoza

Men's personal conduct is the basis from which all of the world's affairs derive. Spinoza's insight was to understand what motivates this conduct and how self-knowledge can help us improve it. He saw that all life endeavours not just to survive but to pursue happiness, and so all behaviour is dominated by desire to have pleasure and to avoid pain: whatever gives us pleasure (directly or indirectly) we call good, and whatever causes pain we call bad, and this is the basis of all human emotion. Critically, because we have innate empathy with other human beings this means we are also influenced by their emotions: it makes us happy when we see them happy, and causes us pain when we see them suffer. This is even stronger if they are happy with us or suffer because of us, because we perceive we are the cause of it. It's because of this we try and make others love what we love and hate what we hate, so our own pleasure is maximised. And so from empathy there is also ambition to please others, benevolence (pleasure at giving others pleasure), and guilt and repentance (which are mostly pain at causing others pain). It's also because of this that people who've done wrong often hate their victim: they feel pain at causing others pain, so they wish to destroy what they see as the source of their pain, which is the person they've wronged. It's from desire of pleasure and empathy with others that there's envy and jealousy: we envy people those things that cause them pleasure because we also want this pleasure. It's also because of empathy that hatred is only increased by hatred, because being hated by another causes us pain, and so we hate in return the person causing us this pain. So hatred is self-perpetuating and can only be destroyed by love. In practice this means hostility, hatred and fear will naturally increase each other, but an attitude of compassion and friend-liness may dispel them. Lastly, it's because of empathy we are most affected by those we are emotionally close to, irrespective of whether we are closely related.

Spinoza also recognized that some emotions don't arise 'naturally' but are formed by education and customs: children are raised to associate particular acts and ideas with pain or pleasure according to whether they were approved or condemned by their parents and society. So they continue to have pleasure or pain from these just by association with past feelings. Superstition is similar: we associate particular things together and so they become connected in mind even though they aren't connected in reality. Because our power is limited and often overwhelmed by uncertain external forces in the world, and because our knowledge of causes and effects is limited, we end up associating hope and fear with things we call 'omens', 'charms', etc. As an old Sufi tale relates, such associations are usually unjustified: "A fool saw an Ass's head on a pole in a garden. Asking, "Why is that there?" He was told: "To avert the evil eye!" To which he replied: "You are the ones with asses' brains, and that's why you have set up an ass's head! When alive it couldn't prevent the blows of its owner's stick. When dead, how can it repel the evil eye?""

The problem of ego

Spinoza saw that the consequence of having a self is that it gives us great pleasure to imagine our own power of acting and it causes us pain to think of our limitations or lack of power. This is the basis of pride: we love praise because it gives us pleasure by increasing our idea of self, and we don't like blame which does the opposite. It's also because of pride that we prefer to imagine our powers of acting or 'virtues', and are envious and jealous of those who have virtues we desire to have. So it is that men envy other men for their virtues, which they "tend to do down wherever possible, delighting in the weakness and shortcomings of others, and delighting in their own virtues, which they embellish wherever possible," and this is why man naturally likes to "pity those who fare badly and envy those who fare well."

It follows from this that true humility is very rare because it's human nature to exaggerate its own self-esteem and strive against thinking less of itself, and to project its own faults onto others wherever possible. And so as Spinoza says it often turns out "those who appear to be the most self-abasing and humble are often in reality the most ambitious and envious." Indeed, in reality self-abasement is just as irrational as pride and just as devoid of virtue, since it's as much a mark of pride and an affected ego to consider oneself worthless as it is to consider one is worth more than one is. The proud man thinks he's above all others and loves flattery because it increases his pleasure, but hates critics because they reduce it. Self-abasement seems opposite, but only comes from the idea of injured pride, and it's often found to be such people that envy others most and are always concerned with the defects and shortcomings of others. It is like those who make a noise of being moral to make others feel worse and make themselves feel better: the most truly moral men are those who do what they think is right without concern for appearing to do so.

It is partly because of pride that there is love of fame: a fondness for pleasure dependent upon the praise and attention of others. The only difference is that the desire for fame rests as much on the pleasure that comes from being desired by other people as it does from the pleasure of having one's pride and 'power of acting' inflated. But it's equally as illusory and unstable. As Spinoza well observed,

Day by day, with anxious care, the person who glories in the esteem of the crowd strives, acts, and experiments in order to preserve his fame. For the mob is variable and inconstant, and so fame, unless it is preserved, quickly vanishes. Indeed, since all vainglorious people desperately desire to gain the applause of the crowd, each one readily plays down the reputation of another, as a result of which there arises an immense desire to crush one another by whatever means, and the person who finally emerges as victor feels more glory for having done harm to another than for having benefited himself. So this glory, and contentment, is indeed vain; for it is valueless.

The key to freedom?

Because we see things causing pleasure of mind or body as good, we naturally desire them, and because we dislike those causing us pain we're naturally averse to them. The result is that unless we cultivate the virtues to control these emotions, they and the desires that spring from them may easily be excessive and become destructive. In other words, as Spinoza taught, people can only attain true contentment if they can learn to govern emotions, rather than be governed by them. For as long as we are determined by our passions we will be determined by the external things that cause them; but as far as we govern ourselves we may be self-determined and free. Indeed, every sage in history has taught the necessity of mastering desire and passion if enlightenment is to be attained, since lack of power in controlling and restraining desires is slavery, for such a man is not his own master but subject to the power of other things and cannot obtain freedom or enlightenment.

The first step to achieving this self-mastery is thus to know oneself, and recognize that many vices such as greed, lust, avarice and ambition, are nothing but self-destructive emotions of uncontrolled love for their objects. It's essential to realize one doesn't make oneself happy by pursuing destructive or harmful desires of greed, anger or hatred. In the grip of passion we wish at all costs to have or crush or destroy the object of our desire, but we begin to free ourselves the moment we pause to consider and understand why we are feeling this way. It's also true however, that understanding alone is not always enough to overcome desires. It has been taught that negative passions should be overcome with positive emotions of compassion, friendliness and love; but whilst these can help, passions do not really have opposites, since we control them by virtues such as temperance, prudence and continence, which are not emotions but powers of the mind by which emotions are controlled. The key is to learn the art of self-control; the art of having an 'ever-present mind'. We should discard the delusions of the ego, realize that possessions are unnecessary, and see that whilst emotions such as hatred, greed, pride, jealousy, anger, are all to a degree perfectly natural consequences of human nature, they are harmful to happiness and must be overcome. We should learn as Spinoza said, "to hate no one, despise no one, deride no one, be angry with no one, and envy no one"; that all attachment and hatred, and all negative emotions arising from them are self-destructive. Self-mastery, and the cultivation of an open, compassionate and kind-hearted virtue are the essential means of achieving this.

It may be observed that we often see what is better, but still choose to do what is worse. That we often ignore our better judgement and do things even when we know the harm it may cause ourselves or others. But the reality is that people don't see the better and choose what is worse; they always choose what they think is best. It's only when you don't really think something is good or don't really understand it that you aren't concerned whether you attain it. The real battle everyone knows about is not really one of good versus evil, or resistance versus temptation; it is of ignorance versus understanding. This doesn't mean harmful instincts don't exist and evil doesn't arise from them. But it means harmful actions spring from ignorance, and when one has the understanding to realize such instincts are harmful one has the motivation to overcome them. As Leibniz observed, "Indifference arises from ignorance, and the wiser one is, the more one is determined to do that which is most perfect." It's only when we are ignorant that we don't care or do harm. When one truly realizes that virtues increase true happiness and vices destroy it, one has no problem trying to cultivate the former and trying to get rid of the latter. For what virtue is there you will not be prepared to pursue, and what vice is there you won't strive to overcome when you realize that the purpose of existence depends upon it?

Chapter 10

The myth of impulsiveness?

There is often a feeling that controlling all desire and emotion in life would restrict all impulsiveness and freedom. But whilst those who act on all their desires and passions consider themselves to be acting from their own power and to be most free, doing what they please; in fact those who act on desires as they happen are least determined by themselves and most determined by the external world as their senses are pulled by one thing after another. To act 'on impulse' is really to be least free, since it is to be least determined by one's conscious self and most determined by other things, as an animal is by other objects. The source of all man's freedom is to think before acting: to pause and consider before doing something. It is this power that distinguishes us from other life, and as Locke observed, "from the not using this right self-control comes all that variety of mistakes, errors, and faults, which we run into in the conduct of our daily lives and our endeavours after happiness." Acting without consideration isn't the way to achieve happiness since not only is it obvious that man is compelled by many conflicting desires and easily prone to desire things in excess and to his own destruction. It's also evident that whilst everyone chooses what they think is good, what appears good often turns out to be the opposite and many things that seem pleasurable don't bring lasting happiness at all.

To be determined by our own judgement is not to restrict freedom: it is to be most free. To choose what is best for ourselves and do it. We may ask if it's possible to control even unconscious desires, but the answer is that all people suffer from conflicting and repressed wishes; this is natural. What is important isn't that we expect immediately to be free of them, but that we learn to be aware of ourselves and our actions. If we don't learn to understand and control our own desires, fears and anxieties, then these will continue to govern our behaviour even when we are unaware of it. The afflictions of the undisciplined and unfettered ego are numerous: pride, arrogance, guilt, anxieties, narcissism and fears all accumulate and take over. Accordingly, one should train the self to be aware of its thoughts and actions, to understand them, control them, and discard those which are detrimental to living. Then gradually, by the attainment of mindfulness and self-government, we can learn to discard hindering desires, fears, and anxieties, and ultimately, be free of them. Few attain mental and emotional self-sufficiency. Instead we transfer or project our emotional desires, attachments and insecurities onto other people and things. But without the attainment of this self-sufficiency we can hardly begin to think of attaining true happiness for ourselves or others.

The burden of responsibility

Even if we understand the consequences of our actions and importance of pursuing the right path, it's not always easy to know which is the right action to take or how best to use our time. One may realize the importance of pursuing virtue and desiring to do what is most perfect, but how are we to know what's best when there are so many options, such little time and so few resources at our disposal? So there is the old complaint of wisdom's futility, that one who has it sees more, but envies those who don't for their contented indifference whilst he suffers every question in difficulty, having the burden of always seeking to do what is best. But ignorant indifference isn't real contentment, and this burden has always been the mark of growing wisdom: real happiness can only consist in being aware of the value of life and its use, and of consciously directing ourselves to this end. It's only when we are aware of what we are doing that life begins to have some value.[25] The lowest form of conduct is to do as we please; the next is to begin pursuing virtue and realize the difficulty that comes with this; and the highest is to have achieved virtue and be free of difficulty; to have endured difficulty in order to realize this virtue.

Yet since as it's said, "the way of tranquillity is narrow and difficult," it's not surprising that we should meet with difficulties on this journey. It's only with dedicated effort that one can learn the skill of spontaneity: of accepting what comes, attending to what is, and of doing what should be done. In the meantime we should just be sure this burden doesn't become a cause of impatience or indecisiveness. Growing angry with existence and impatient with others because of the difficulties of life is the mark of egotistical impotence not wisdom, and a wise man should become more patient and humble the more difficult life becomes. And whilst pursuing this end, when the way seems hard to find, when others seem clear-headed and one can't understand, when all else has purpose and one can't find one's own, one should only remember what it is one strives for and persist with one's practice. When difficulties come, "darkness of soul, turmoil of spirit, inclination to what is low and earthly, restlessness arising from many disturbances," we should keep fast with our perseverance, holding firm in resolution, and remember that there is no preparation for later. If we are angry or dishonest or hateful in this moment, this moment is ruined by our misconduct. Therefore we should strive in every moment to create and preserve happiness for ourselves and others.[26]

Discarding anxiety

Once we realize how much depends on each decision it's easy to become plagued by anxiety and indecision. Yet a life dominated by anxiety is worth nothing at all, as Epicurus taught. And whilst it's rarely easy to choose the right course, indecision is as much a vice as failure to consider properly at all; because whilst the latter usually yields wrong action the former yields no action or action that is too late. We should consider everything we do and exercise all our powers to know and do that which is most perfect, but we shouldn't allow consideration to become indecision.

When faced with many options we should always recall that most difficult questions in life do not have easy answers and the best we can hope to do is to give a matter due consideration, ask the advice of others, and do what seems best. The wise attitude is to always retain the calm and balanced reflection that all a man can do is what he thinks is right at the time. If we have judged with honesty, temperance and prudence in mind, and good-will in heart, then we cannot expect to do any more. If an action turn out well, then all the better; if it does not, then so be it. Resoluteness and decisiveness, but not rashness, are essential qualities that must be cultivated by one who seeks 'The Way'.

And in the meantime, we shouldn't forget to retain a sense of humour and a balance of perspective in all things. The only thing is knowing where the balance lies: knowing that good humour and light-heartedness are essential but life is not a joke, and that whilst enjoyment is necessary who has time to waste with trivialities? As it is said, "This day will not come again. Each moment is worth a priceless gem."

Chapter 11

Passionate illusions

Even though it's necessary to master passion and desire, it's also inherent in men to desire women overpoweringly and above all else. Indeed, given reproduction is the basis of existence and self-preservation is the strongest desire in man, it isn't surprising this desire is hard to conquer and we find the highest to the lowest minds governed by the "desire of desires." Yet lack of control in this respect is one of the great sources of life's difficulties: of all the misuse, infidelity, dissipation, guilt, and self-destruction that derive from uncontrolled sexual desire. One of the most necessary disciplines a man may learn is to control this desire and sublimate this energy into fulfilling life rather than dissipating it. Without such discipline this desire may become totally opposed to the mind's control, and without some control there cannot be happiness. Within the right context sex is a good and necessary thing, but promiscuity and faithlessness are just the subjugation of mind to desire: it's only through fidelity and self-discipline that a man may begin to understand himself and others, and achieve that happiness which comes from true freedom.

Yet it isn't necessary to despise oneself or one's desire to achieve this: we do not conquer one passion by the force of another. We do so by being motivated to realize the purpose for which we are created, and by cultivating the virtue of mind to achieve it. Continence just requires resolution and mindfulness in conduct. All desire and all pain ultimately derive from the mind; if a man truly wishes it he may conquer any desire and achieve what he will. People convince themselves that desires are uncontrollable but this isn't the case at all. Pain and desire just derive from the mind; love and hatred just derive from the mind. Passion, anxieties, fear, all derive from the mind. Therefore learn to know your mind and you'll see that, as it has been said, "Nothing outside yourself can cause you any trouble." Many thoughts cross our mind but we are free to choose those we pursue and those we leave behind. If we get angry, or anxious, or in a passion, it's not because of other people or things; it's because of our own mind. With all desires it's possible to learn self-control and mastery if we wish. The fool regards each man as a machine driven impotently by the force of his genes, but the wise man sees them for what they are: tools placed at the disposal of each human mind to express itself as it will.

Resolving contradictions

The difference between men and women in essence is that men are driven to desire women and women are driven to desire being desired. The result is that men are naturally inclined to promiscuity whilst women are naturally inclined to be faithful, as long as the fidelity is returned and they continue to feel desired. This follows from evolution: men can happily propagate their genes without committing anything upfront, whilst women must endure a long pregnancy and the burden of a helpless child for many years. So women aren't blindly attracted to the opposite sex in the same way as men; they're more concerned to secure a committed and worthwhile partner, and so they are usually content to desire a single man and be desired and loved by such a man. And thus whilst a man's choice of partner is often dominated by superficial attraction and physical appearance, a woman will consider assets like power and wealth to be equally important since these are usually a better mark of a strong mate and successful partner rather than whether he happens to be born with good looks.

Yet men of course also wish to be desired and take a strong interest in their children, and women also have sexual desire, and whilst it might seem their desires contradict, men and women both ultimately desire the same thing: the creation of a family and happiness from this. Problems only arise because people fail to distinguish between the illusion of desire and its reality. The illusion of promiscuity is a conquest of pleasure, but the reality is dissipation, emptiness and unhappiness. Because men desire women they think the more they pursue this desire the happier they'll be, but in fact the opposite ends up being true. Man being man cannot help but desire women, but the perfection of love is that a man masters this desire and binds himself to a woman and remains faithful to her. Many men naturally choose to indulge their desires without commitment, but if a man cannot discipline himself to remain faithful to one woman, he will never find himself satisfied by a thousand either. Likewise it's easy to turn a man's head but it's more difficult to capture his heart and mind; yet if a woman wishes to have the happiness that comes from being loved instead of just desired then it's necessary to do so. This means learning intentions and judging character before deciding whether to commit, rather than committing first and judging afterwards.

All the incessant striving and endless pursuit of sexual desire is just about creating the next generation; as Schopenhauer observed, " ... all guided by a delusion that conceals the service of the species under the mask of an egotistical end." To approach life without seriousness and have many casual relationships isn't the way to happiness or fulfillment. It may be hard, but those who seek happiness and wisdom should learn to practice restraint and self-discipline irrespective. As it is said in Proverbs, "Reserve it for the woman you marry. Don't share it with strangers." The true way that is intended for men and women is the greatest happiness they can attain: the union of man and wife and the love and creation of life that comes from this; of raising children within a family. It is also in this context that one can learn many of the essential practices for achieving wisdom: learning the need of compromise in a relationship, of taking into consideration another's feelings and thoughts as well as one's own, of being responsible for others and working for them, and learning to control and direct oneself to this end.

Understanding and communication

Solitude has its use but being a responsible member of one's society and one's family are one's divine as well as worldly duties. The two essential ingredients for any successful relationship are that firstly we shouldn't take what we have for granted: we should be appreciative for what we have and work to keep it. And secondly we should remember to communicate, since people can't read each other's minds but only know and understand what others think and feel by what's communicated to them. This means we should both learn to understand ourselves and say what we really mean, since when we don't understand ourselves or express ourselves clearly a great deal of heartache and trouble can follow. It also means we should learn to listen and to see things from others' perspective, since we often become upset, angry, or misunderstand others because we only see situations from our own perspective. Learning to see from others' perspective is the beginning of understanding why others act as they do, and tempering our own ideas and emotions with those of others.

The reality is that true love requires as much discipline, faithfulness and trust as desire and attraction. It doesn't come freely or easily but is cultivated by effort as well as desire. This means even though it's natural to be attracted by appearances we shouldn't be led to chase people according to how they appear: we should learn to love them for who they really are. And just because there is no such thing as a trouble-free person, or a perfect love at first sight, or an ideal and effortless relationship to find, doesn't reduce the magic of the world or its romance. There is still love to be found, only its perfection is different from the way we anticipate and we must work to maintain it. All relationships must endure difficulties, but the most perfect union is the one that withstands the troubles of circumstance and the test of time. For love is not merely in what one desires but also in this: in consciously bending one's will and one's desire to accord with the best interests of the person one loves. Life cannot be all as we wish it, and it requires many sacrifices; but it is by these sacrifices that we learn that life can be much more rewarding and worthwhile than our imagination.

Chapter 12

Healthy body, healthy mind?

It was once a habit to think of the body as "flesh clogging the immortal soul" and as impurity dragging down the pure mind. But mind is no purer than body; we should purify each equally and when they are in a state of health and harmony the body becomes an extension of the mind, not a hindrance to it. Indeed, not only are physical discipline and the pursuit of good health a powerful means of learning to discipline and control one's self, but while we exist on earth our happiness and contentment depend heavily on the state and condition of our bodies. Appropriate physical exercise is the most proficient means of releasing stress, generating internal energy, promoting the unity of body and mind and learning to be relaxed and at ease with oneself. It isn't necessary that we should expect to reach the peak of physical perfection, but it is within the grasp of everyone to ensure they exercise regularly, eat well, and consider the body as more than the 'temple of the mind', but also the means by which it lives, thinks and breathes.

We often take it for granted that if there's nothing wrong with us and we aren't in pain we need do nothing more. But if we don't look after the body and keep it healthy then poor health will set in, and neglect only ends up yielding a body unfit to house the mind that governs it. Just as the mind affects the body, the body affects the mind: if it is weighed down with torpor and unhealth then so is the mind; if it be sick, then so are we; and if it's disturbed so is our state of mind. Emotional stability, mental clarity, anxiety, concentration and sleep all depend partly upon the state of one's health, and without stability in the body it's difficult to attain stability in the mind. But with correct physical discipline comes temperance, concentration, and discipline in sleeping, eating and character.

Yet whilst this might imply happiness is beyond the grasp of those who suffer disease or disability, impairment of the body, just like that of the mind, is an impediment to happiness, but it won't prevent one who wishes to achieve it from doing so. What determines our state is what we do with the circumstances we find ourselves in. It's mostly beyond the means of someone born with disability or disease to determine their physical health. But they can still cultivate mental and spiritual health and achieve happiness if they do as much as they're able.

It isn't about whether you win

Whilst the benefits of being healthy are apparent, it's also obvious that many who undertake exercise are themselves examples of how we shouldn't behave; lacking virtue, emotionally unstable, arrogant and vain. But all this shows is that one should undertake the right kind of exercise and with the right intention. The right kind of exercise is one that aims at the health and well-being of the whole body and mind: to increase their strength, vitality and balance. And the right intention is to improve one's entire self and be healthy in mind and body; not merely to improve appearance or achieve victory over others. It is the delusion of pride that it must attain victory in everything and excel over others. Our only concern in life should be to participate with as much skill and effort as we are able: to make our brief existence an expression of every inch of our being and potential. Of what concern is it whether we win or lose? We shouldn't be concerned to 'beat' others, only to do the best that we're able, and if others do better we should be happy for them.

Pursuing physical discipline does not supplant the need to practice virtue. Such discipline is an asset to virtue, but without virtue it is not only useless but damaging since one will find one has more energy to feed the unsettled mind and its desires. We always seek to master and control what is around us: the exterior world, its possessions, and other people. But the only path to happiness is through mastery of oneself. All else proceeds from this: mastery of self, desire, and the body and mind. Many suffer by projecting their unhappiness and problems onto their bodies; but people can't be happy when they are trying to alter themselves in order to accord with what they think they should appear to be. A wise person makes their physical practice their spiritual practice: the preservation and fulfillment of their life and health, the creation of energy and happiness, and the yoking of all their energies to this pursuit.[27]

The illusion of intoxication

Whilst it's useful to relax, it's essential to realize that it isn't necessary to use or consume any substance to unwind or 'be oneself', or to interact with others. The true joy of life doesn't need anything external and anyone can learn to be free and relaxed without the need for anything but the power of their own mind. People convince themselves they're more at ease when intoxicated, but the reality is that the excessive use of alcohol destroys health, disrupts the mind, creates indiscipline of character, and incapacitates the power of the mind to discern what is good for itself and others. Alcohol gives the illusion of self-confidence and well-being, but the reality of drunkenness is unhealth, incontinence and irresponsibility: in speech, in desire and in action. If we wish to attain strength of mind and character we should practice moderation and realize that clarity and some self-control are essential to a mind that seeks happiness. The only usefulness of alcohol is if it's consumed moderately and doesn't cause one to lose control, and its only benefits are contingent on this; with regularity and dependence come deterioration and the disunity of body and mind.

The same isn't true for narcotics however, for not only are their effects more powerful because they act directly on the nervous system, they also cause greater destruction because of the force of this effect, corroding the system that is the very foundation of mind and body. Drugs simulate a state of well-being, ease, relaxation and energy, but their reality is the very opposite: the destruction of the body's health, the clouding of the mind's clarity, the disturbance and corrosion of its equilibrium, and the weakness of addiction. There are those who think that they improve their creativity and imagination by the use of such substances, but drugs impair mental activity, they do not improve it, and creative thought is best undertaken by a strong, free, and stable mind, not one labouring under the yoke and delusion of addiction. All people suffer from anxieties, depressions, and uncertainties; but it isn't by means of external substances that one can be freed from these. It's only through the practice of mental and physical discipline that people can learn to manage their lives more effectively and wisely, and to develop a free and relaxed inner tranquillity and contentment that is not easily disturbed by the stresses of the outside world.

Chapter 13

The use and limitations of intelligence

It might seem great intelligence is necessary for achieving the highest wisdom, and since few are born with such intellect the happiness that comes from having the highest wisdom must be beyond most of us. But intelligence isn't the same as wisdom and true happiness doesn't require more intelligence than anyone possesses or can develop. Intelligence is only required insofar as each person should order themselves and their life according to the fundamental principles of human wisdom, and these aren't difficult to grasp. Everyone should think and study for themselves and pursue wisdom and happiness, which are in the pursuit of some rational understanding, the improvement of oneself, the pursuit of ethical conduct, the acquitting of one's duties on earth, and the right use of the life one has been given. No great intellect is needed to know these things, and in determining wisdom right intention and right effort are often more important than intelligence, since intelligence can complement these but it cannot replace them. There are many clever people who aren't wise at all and we shouldn't think that by possessing intellect or accumulating knowledge we'll acquire wisdom: possession of knowledge isn't equivalent to being wise, and learning by rote isn't a substitute for understanding. This isn't to say knowledge isn't necessary, since some understanding of oneself and the world depend on it. But knowledge and intellect alone aren't enough: wisdom depends on how we use and apply them. Intelligence can lead to wisdom, but it can also lead away from it.

People all around the world persevere. They expend great effort and determination in living their lives and pursuing the ends they desire or think are right. But it's essential we shouldn't just become trained in persevering after something. We should also become trained in thinking about what we are persevering after: the real purpose of life. This is the real use of intellect and learning to think for ourselves. Intellectual pursuance is only necessary to everyone to the extent that we should found our life and works on truths rather than false principles; since tenacity in principles and purpose are admirable in a man, but when aimed at the wrong end or founded on the wrong basis it is plain how much harm they may do.

The difference between eloquence and excellence

Just as intelligence and wisdom aren't the same, reputation and eloquence may complement wisdom but they aren't substitutes for it. There are many who dress up their ideas in the most complex and confusing language they can conjure up, but the most truly intelligent people are those who can impart important concepts without complicating or confusing them. Yet since impenetrable eloquence never fails to impress, there will always be those who invent meaningless words and obfuscate to amaze, and "never actually benefit the world for all their profound wisdom and intricate disputes about unintelligible terms." As it is said by the Sufis, "Those who know little, and should be studying rather than teaching, like to create an air of mystery. They encourage rumours about themselves and pretend to do things for some secret reason. They always strive to increase the sense of mystery. But this is mystery for itself, not as the outer sign of inner knowledge. People who know the inner secrets generally look and behave like ordinary people."

The true power of words is when they're spoken from the heart, and true teachers are those who can convey with simplicity that which needs to be conveyed. Undue concern for eloquence or appearance yield nothing substantial, and undue concern for reputation is nothing but desire for fame. Those relying on witty eloquence in place of real virtue are like "rusty boats beautifully painted; if you set them on dry ground they look pretty, but once they go into the sea and oceans, into the wind and waves, they are in great danger," as a Zen master warned. A true student shouldn't be concerned to think about being clever or having the appearance of great intellect; they should just be concerned to penetrate the heart of the matter. The difference between not understanding something because it's difficult and not understanding because it's nonsense is not always straightforward. But most often if something has truth it may be said plainly and then it will be understood plainly. This doesn't mean the path to grasping the most important concepts in life is easy or straightforward. But it means we should never be afraid of some respectful questioning when we encounter something we don't understand. And if we still cannot understand what is being said or see the evidence for it, we should learn to reserve our judgement and merely say we cannot understand it until we can see if it has any meaning or not.

One reality?

Since the dawn of mankind there have been many wise and enlightened teachers, and they have been of all races and of all faiths, some orthodox and some unorthodox, but all original and all perceiving that the attainment of a higher state depends upon the refinement of man and pursuing some realization of the true nature of this reality. There have been many different forms of teaching, and many different sects and doctrines and practices, in many different parts of the world held by many different people. Yet there is only one reality that lies beneath these in which all people share, and one can very well adhere to the form of one's own path and one's own culture and religion, whilst recognizing that ultimately all different religions are but different facets of a single diamond; that all are spokes that lead to a single hub.

Religions are institutions developed to help men improve themselves and become closer to God; to realize that the outward is the path to the inward. But all religions and spiritual teachings are only outward means of leading people to the same inward truth. As it has been said by the Sufis, "religion is a vehicle; its expressions, rituals, moral and other teachings are designed to cause certain elevating effects, at a certain time, upon certain communities. ... but for those who don't understand the means always becomes the end; the vehicle becomes the idol." Religion doesn't define God and God isn't defined by any particular religion. The problem is form and substance become confused. People get so stuck on forms they fail to perceive the substance that lies beneath them: the truth they're intended to impart. Practices become ritualised and imitative, concern with external form arises; what is good and true becomes diluted and mixed with the false and superficial. People begin to idolize teachings and teachers, to mistake scholasticism for wisdom, and lapse into doctrine and dogma. And this is the only thing one should avoid; becoming caught up in unnecessary learning and lost in useless subtleties and distinctions that have no use in true understanding or self-improvement.

The true pursuit of wisdom is, "to hear much, to pick out what is good, and to follow it," and so it has been said: "The sacred books of the world are for all to read. They are not meant for the members of one faith alone ... so the seeker absorbs things in other faiths which will enable him to appreciate and understand his own faith better." A wise man learns to examine all ways and to understand the truth that underpins all great faiths, without needing to abandon his own and without getting caught up in unnecessary concern for outward detail. There is no particular path that leads to God. Any path that encourages men to have faith in God wisely, to be good in conduct and to seek true happiness for themselves and others, shows the same way. As Master Eckhart wisely taught, "Each person is given their own path, so if you see or hear of a good person following a different way from yours then don't think it is a wrong path, for not everyone follows the same way, but the benefits that one way gives are also present in others."

The means to attain the great truth and joy of life are possessed by everybody. No great intellect and no library of obscure philosophies are required to attain it. One should absorb and pursue teachings in order to understand, progress and improve oneself, but when this improvement is attained the only teaching we need is in our own minds. What lies within is perfectly sufficient, if we can learn to use it correctly. This is the meaning of Rumi's words when he says: "Mankind passes through three stages. First he worships anything: man, woman, money, children, earth, and stones. Then when he has progressed a little further, he worships God. Finally, he does not say: "I worship God" or "I do not worship God."" The reason he doesn't say it is because it's obvious: God isn't there to be worshipped. God permeates every facet of reality and existence. Therefore what need is there to say you worship It? Your worship is evident in how you live and conduct yourself, and there is no need to say anything further.

Chapter 14

The difference between equanimity and indifference

Many teachers have taught it is necessary to overcome desire, and that man should learn to detach himself from this world and its impermanent nature. Many ancient philosophers imagined a state of perfection as one completely free from desire and all trace of the mortal condition. But there's a danger that in aiming for this detachment and seeking to escape the turmoil and intransience of life one ends up rejecting the world and attachment to life within it. Because we've sought to find a way to avoid death, to be free from desire, and to escape this reality for a higher realm, some have ended up pursuing detachment to the extent that they have become so indifferent to the world that they don't care about it.

But the ideal end isn't to be undisturbed by the fate of those around us, with a care only for our own salvation. And we shouldn't pursue the spiritual path to escape from this world; we should do so in order to perceive and enjoy it all the more fully. We shouldn't practice self-discipline in order to become oppressive and dogmatic, or to suppress the joy of life in zeal and anxiety. We should practice to achieve happiness with tranquillity, freedom, directness and spontaneity. Some have made the mistake of considering the means to achieving transcendence is becoming indifferent to all dualism in the phenomenal realm, but to have equanimity isn't to be indifferent, and fulfillment can't be attained by shirking responsibilities. We can't transcend the reality of emotional connections, or old age, death, and suffering: these are part of the nature of human existence, and retreating or detaching from the world doesn't alter them.

It's been a tradition for some monks to meditate on the human body as blood, flesh, pus and hair in order to overcome desire, or to meditate on this world as impermanence and suffering, or to visualize some other realm of infinite happiness and eternity in order to overcome attachment to the world. Others have meditated on old age as the destroyer of beauty and strength, or on the inevitable reality of death, or the passing away and decomposition of all that lives into rotting matter. But the real use of such methods isn't to generate a disgust for the world: it is just to recognize that the things of this world are impermanent, to learn detachment, to discard pride, and to have equanimity. As the old saying goes: "People are tied down by a sense object when they cover it with unreal imaginations; likewise they are liberated from it when they see it as it really is." Generating contempt or hatred for the world, or a view that all life is affliction, is only replacing one passion of desire for another of aversion; which is not an advancement but a regression. Making oneself hate something because one cannot control one's love for it is no better than failing to control oneself at all. The only use of such contemplations is to perceive what is of importance in this life, which are not the appearances of things or objects, which are often illusory and uncertain, but how we live our lives in relation to others and the world, and to learn some equanimity in our approach: to set aside the ephemeral and to help us realize that clear vision which sees all things as one.

The perfect state of man isn't to be free of all desire or completely untouched by happiness and sadness. It is to be in control of oneself, to appreciate the joy of life, to share happiness with those around, to have eliminated negative states of mind and body, and to have attained a state of equanimity: to be ready to depart whenever one's time is due. And the true use and virtue of stoicism is when one is stoical in one's attitude to oneself and one's own fortune; not when one is quick to be stoical with others, or the fortune of others. It isn't emptiness of concern for others that is needed; that is indifference. True equanimity is emptiness of self.

The practical use of equanimity

We might still ask what equanimity really means. How is it possible to both remain the same and care about others? If you're unaffected by them you must be indifferent; but if you're affected you don't have equanimity. How can you be tranquil if your heart is tied to the world and those within it?

But the answer is that to have equanimity is to be detached and not easily affected by the things of the world, but it isn't to be indifferent to those in it or unable to adapt to the change within it. In practical terms this means regarding all people equally and "receiving guests with the same attitude as when alone, and having the same attitude alone as when receiving guests"; having the calmness of mind that comes from having evenness toward all and behaving equally well toward all. Yet treating all people equally doesn't mean treating everyone the same. We should treat everyone with understanding and compassion but that doesn't mean we'll treat a violent thief the same way we treat an old friend. We should merely act when appropriate in the appropriate way. Having wisdom means adapting to circumstances, not being confined by them. Being free from constraint but not from care. It is only by the attainment of this equanimity that one can learn to be "free everywhere, at odds with no-one, and content with this or that."

What is really meant by detachment is learning to control our desire for other things such that we govern our desires and aren't governed by them. Thus one shouldn't be free from all love and care; from all concern and emotion. But nor should it be wild and uncontrolled. When it's said: "Be free of distractions and attachments," it doesn't mean cut yourself off from the world. It means don't let yourself become distracted on your path to achieving wisdom, and don't let yourself become attached to harmful things that may detract from it. "Eyes that aren't attached to form are the gates of wisdom. Ears that aren't attached to sound are the gates of wisdom."[28] And in respect of the world equanimity just means tempering all attachment to things with the knowledge that they aren't permanent, but all will be separated from us sooner or later.

An ascetic mind

It was once thought if you really wanted to seek God you should abandon everything and quit society, since as long as you remained in it you'd be distracted by worldly things and tied down by the commitments of ordinary life. But anyone can seek spiritual truth whether they're in society or not and it isn't necessary to go looking in the mountains to find it. The ideal world wouldn't be overrun with flea-ridden hermits, and whilst happiness doesn't necessarily need other people they can certainly assist it. In the old times monks used to roam freely without attachment. But it doesn't matter whether you roam or have a fixed abode; it only matters that your mind roams freely. When the mind roams freely in its quest for God what concern is there for physical location? The true way is just to ensure everything you pursue is with the truth in mind. As one Christian master taught; "I do not condemn those who wear fine clothes or eat well, as long as they have love. I do not regard myself as being any better when my life is demanding than when I see that there is more love in me."[29]

A master isn't one who must withdraw from the world because they're unable to remain mindful in it; it's one who can learn to do this wherever they are, and whether in society or not. As Master Eckhart taught,

Some people like to withdraw from company and prefer to be alone. That is where they find peace, when they enter a church. I was asked is this the best thing? My answer was 'No!' and this is why: That person who is in the right state of mind, is so regardless of where they are and who they are with, while those who are in the wrong state of mind will find this to be the case wherever they are and whoever they are with. Those who are rightly disposed have God with them. No one can obstruct such a person because they possess God alone, intend God alone, and all things become God for them. Thus they are at peace in all places and with all people.

And thus the Zen poem says:

> He has established himself on a mountain,
> So he has no work to do.
> A man should be in the market place,
> While still working with true reality.

One should learn to master oneself wherever one is and learn to live according to virtue in all places and circumstances. Solitude has its use in contemplation and meditation, but it's by means of this one learns to engage more properly with the world, not retreat into oneself alone. We cannot help but be born into society, and we cannot do without it, and if we wish to be happy in it we should take an interest in it: for society, and the politics which govern it, are the means and the context in which all are facilitated in their struggle for happiness.[30] The true expression of pursuance is to both strive for mindfulness and sincerity inside, and to work and help others outside.

Chapter 15

What is worship?

We are told to pray, build temples, light incense, observe fasts, make pilgrimages, and all such things in order to encourage us to pursue the spiritual and to be disciplined in our pursuit: to cultivate the right attitude and state of mind within, and let the external express the internal. But what matters is what we cultivate within ourselves; our purity of mind and actions. What benefit is worshipping statues or performing rituals if we're just concerned to get something from it? And what benefit is going to pray if we're just concerned for our own salvation? The true use of prayer is when we carry its illumination within. True prayer is being mindful and observing precepts in whatever we do, constantly cultivating virtue; not just reciting spiritual texts when it's time to pray. And true spiritual practice isn't determined by just adhering to rituals and practices; it's determined by what we do and think at all times. It's been wisely said, "To say whether you believe in God or whether you don't believe is not important at all."[31] One who promotes rational conduct, self-criticism, tolerance and compassion but calls himself an atheist has far more religion in him than one who spreads ignorance, bigotry and hatred but claims to be a teacher of religion. One who lives virtuously and helps others but says he doesn't believe in God has more faith in him than one who visits the temple every day for his own selfish gain.

Rituals and pilgrimages start and end; you start them, finish them, and then get on with your daily life. But true practice doesn't start or end. Properly speaking, life is our practice, and if we live it poorly our practice is poorly. There are no breaks from this practice: reality doesn't start when we sit to pray and finish when we get up. It is always present. Thus it is said, "One who keeps the gates of his senses pure, his body and mind still, inside and outside clean, builds a monastery ... If you can wipe out evil desires and harbour good thoughts, even if nothing shows, it's worship." Real worship is invoking God in your thoughts and actions, overcoming greed, anger, and hatred, and cultivating virtue and awareness. And real pilgrimage is travelling along 'The Way' wherever you are. Making the great journey to improve yourself: to find virtue and realize the truth. One who attains this enlightenment makes the great pilgrimage even if he never leaves his home town.

Training the mind

The pursuit of good conduct, working hard, looking after health, the company of family and friends, and remembering God can yield great happiness. But the mind, like the body, forgets things if they aren't practiced regularly: we must work for those things we wish to have and to keep those things we already have. If we really want to obtain a certain kind of wisdom we must train the mind to achieve some self-control and realization: to achieve the mastery of oneself, unify one's spirit, and place one's mind in the frame of "perceiving all things from a perspective of eternity."[32] As the proverb says, "When in poverty people take care to work with discipline for virtue and prosperity; and when in prosperity their carelessness and indulgence makes for disaster."

The true uses and effects of prayer or meditation are pursuit of the supramundane, and from this the improvement of wisdom and calmness in the realm of the mundane: in the everyday world. The spiritual use is that one gains an abundant and inexpressible inward delight in existence. The practical use is that having a calm and disciplined state of mind allows us to be more awake, attentive and useful in daily life. The uncultivated mind and attention are often scattered and distracted, lacking focus, and limited to the narrow. But practice can bring clarity and originality of thought, directness of purpose, and a sharp, serene and well-focused mind that is wide open.

It is said in Zen it takes only a minute to learn the secret of meditation: *to sit and keep the mind free from all thought*, but one will be lucky to accomplish it in a lifetime of practice. "Thinking of not thinking of anything at all. How is one to think of not thinking of anything at all? Be without thoughts, this is the secret of meditation"; or as Plotinus said, "Cut away everything." In everyday reality it's essential to reason and understand the nature of the world; but you can't reason your way to understanding the whole of reality. In meditation one aims to silence the mind and experience experience.[33] Thus one must stop all the activities of the faculties, and hence the Zen teaching that "if a conception arises, cut it off in the twinkling of an eye." So an enlightened master said, "However deep one's knowledge of philosophy or science, it is like a piece of straw flying in the vastness of space." This is the real realization I know nothing; that even the little I know is itself nothing. As it was expressed by Master Eckhart:

You must silence all the faculties; the memory, will, reason, senses, imagination, since they only serve to diversify and divide you. ... If this realization is to arrive it must come solely from within, from God. ... Neither the skills of all creatures, nor your own wisdom nor the whole extent of your knowledge can bring you to the point that you have a divine knowledge of God. If you wish to know God then your knowing must become a state of pure unknowing. If God is to shine divinely within you, your natural light cannot assist in this process but must become a pure nothingness. ... The more you are empty of self and freed from the knowledge of objects the closer you come to Him. The very best thing you can do is remain still for as long as possible.

Of course the ultimate purpose of spiritual practice is to gain something; but in practice one shouldn't have any idea of practicing for the sake of something to be gained. One should just "sit wholeheartedly and pay silent attention to the great matter." The essence of enlightenment lies in realizing that it doesn't consist in gaining anything for oneself, but that it consists in giving up the idea of oneself and practicing for the sake of quietening one's mind, serving God, and being more awake and useful to others. This needs patience and perseverance. Only gradually may one surmount inability to concentrate and doubt at the task at hand. As it is said in the Bhagavad Gita "When all selfish desires are abandoned and the mind completely withdraws the senses within, then by disciplined intelligence, thinking of nothing else, the mind gradually becomes silent and absorbed in the Spirit." And hence it is said in the Laozi:

Without stirring abroad
One can know the whole world
Without looking out of the window
One can see the way of heaven.

Transcending duality and realizing emptiness

The ancient sages talk of attaining a state beyond duality; of being indifferent to virtue and vice, pleasure and pain, free of desire and passion, and beyond cause and effect. And this seems confusing, since as long as we exist it seems that things like good and evil, self and other, must always exist, and that a good man should be choosing good instead of being indifferent to it. But "beyond duality" doesn't mean not caring about good or evil, or acting like an idiot without cause. It just refers to the state attained in meditation where there is the cessation of thought. And in everyday reality, whilst it may be said there is no vice or virtue for a master because wrong paths become non-existent to an enlightened man, this doesn't mean virtue or vice don't exist for him. Virtue and vice exist for him just the same as anyone else, and one can only attain the state beyond good and bad when one has cultivated moral effort such that only good remains. This is how one transcends good and bad: there is no other way. In the mundane world, the only motivation of one's actions is to do good for oneself and others.

Likewise the saying that "in reality there's no self" seems a contradiction, since as long as we exist it seems there must be a self and a duality: a distinction between a perceiving mind and what it perceives. But all this means is being enlightened enough to perceive beyond the veneer of ego and pride, beyond 'I' and 'mine'. When scriptures speak of the true end of meditation as nothingness, thinking of nothing, and having a mind that abides nowhere, they don't mean nothing exists. No-thing doesn't mean nothing, and emptiness doesn't mean nihilism. It means nothing exists except in its relationship of subjectivity and objectivity. Ordinarily things seem divided between self and other, but ultimately, there is no division: *real nature and your perception of it are one*. Realizing emptiness doesn't mean realizing life is empty: it means perceiving the absolute. One can learn "the mind is only the organ of thought and the soul itself is one with the infinite." As Hume observed, if you look into your mind you just find perceptions of experiences, ideas and memories; where is the self? And as Kant said, 'I' only derives from the necessity of having a point of view of the world; it is just the perspective from which we view the world.

Saying "emptiness is the ultimate nature of reality" doesn't mean all is empty or the world isn't real. It means the realization that ultimately the concepts of ego and self are empty, that phenomena are empty, and that ideas of duality are empty: mind and matter, subject and object, time and space, all disappear. Insight into emptiness is this realization of non-duality; the mind becomes absorbed in emptiness "like water being poured into water." As Zen master Huang Po taught:

This spiritually enlightening nature is without beginning, as ancient as the void, subject neither to birth nor destruction, neither existing nor not existing, neither impure nor pure, occupying no space, having neither inside nor outside, size nor form, colour nor sound. It cannot be looked for or sought, comprehended by wisdom or knowledge, explained in words or contacted materially. Everything is just the one Mind.

Many spend their lives seeking for enlightenment in something outside; chasing after external salvation. But as master Hui Neng said, "What for do you look outside? There is no Buddha in the mountain. Buddha is in the mind's heart. If your mind is calm and understanding, that indeed is Buddha; if you awaken to this mind you'll become Buddha. You needn't look elsewhere."

At the same time in mundane reality one can't transcend duality, and there's no need to. In the present, in the mundane reality in which our daily lives exist, of course things have independent existence, and the need to distinguish between self and non-self and between internal and external realities is essential to normal life, since a failure to do so is a mark of psychosis, not insight. And it's our actions in this realm that determine whether we obtain happiness and enlightenment, since it's only by striving toward virtue in daily life that one learns what true wisdom consists in. Enlightenment doesn't lie in self-absorption, and grace doesn't come to those just seeking salvation for themselves. The use of meditation is to discipline and open the mind, but as long as we live on this earth we should never cease striving and working for the good and justice of others. As it's said in the Upanishads:

Not by abstaining from action does man attain freedom from action, and not by mere renunciation does he attain supreme perfection. For not even for a moment can a man be without action. All actions take place in time by the interweaving of the forces of nature; the man lost in selfish delusion thinks that he himself is the actor, but great is the man who, free from attachments, and with a mind ruling its powers in harmony, works on the path of true virtue.

Philosophy and religion have shared a view of reality little changed for thousands of years: the material world is a transient and phenomenal reality whose perception is conditioned by the human apparatus for perceiving it, whilst the ultimate reality that lies behind it is mostly hidden and unknowable. If we want to think about this ultimate reality conceptualising thought is useless: "We must leave behind all conceptions of the divine."[34] We might think we should 'feel' our way to the ultimate truth. But feeling isn't the answer either; emotional ecstasy is not coming closer to God. We should go beyond thought and feeling. Ultimate appreciation of existence is partly rational, partly emotional and partly neither: a silent and direct apprehension of experience.

Chapter 16

Freedom of thought

A good teacher doesn't stick to the same doctrine and format in every case, he adapts them to communicate with his audience, since the same method of communication may not be suitable for all places and people. Likewise, a master would employ whatever behaviour or method was appropriate to improve and enlighten his students. "The Way may be through a drop of water. It may equally be through a complex prescription," as the Sufis say; which is the case depends on the circumstances. A good student is not concerned with the form of the vehicle, but only with grasping what it is intended to impart, and a good teacher is not concerned for errors in small things; being broadminded, he is only concerned for the adherence to truth in things that matter.

In spiritual pursuit, learning and study is as a long ladder one must climb up before one can kick it away. This doesn't mean learning is unnecessary, but it means one should see that words and explanations are, ultimately, superfluous. "Description is useless because it tries to circumscribe."[35] In the end, one should discard all ideas; only then can the mind be completely free and unobstructed. But discarding all ideas doesn't mean having the mind of an idiot. It means not clinging to appearances or attachments, "seeing your own nature, knowing your own mind"; and this can't be expressed in words.

Learning to have this freedom in daily life means not resisting change, but moving with it and remembering that the mind is created for learning. Secondly it means not getting stuck on beliefs or preconceptions: being open minded. And thirdly it means turning obstacles into opportunities: not getting frustrated with them but using them as an exercise in adaptation or patience. Life is our tuition. We shouldn't fight circumstances but accept what's beyond our power and use them as best we can; not resisting change but recognizing it as the nature of existence.

It has long been the battle of all kinds of institutions to resist change and suppress truth if it threatens vested power and authority; but as history also shows only those that can adapt and embrace change are those that survive. The right attitude is a deep personal faith, an ethic that relies on principles not rules, and a disregard for excessive formality. It is no secret that religions often struggle to avoid becoming the opposite: prescriptive, dogmatic, oppressive to free thinking, and unable to embrace change because it challenges vested interest.

Who's enlightened?

In practice it's not always straightforward to distinguish true wisdom from pretence or delusion, since there are many who think they have the answer, but unless you know what it is yourself how are you to know any better? But the first thing to realize is that it's not for oneself to judge whether one has attained enlightenment: wisdom doesn't depend on self-assessment, and one who has attained some wisdom not only knows he has still not attained anything at all, but behaves accordingly. Likewise, imitation isn't a substitute for under-standing or self-improvement: there's a great difference between thinking we know the truth and realizing it; between imitating truth and expressing it. Imitation is the first stage of beginning to improve, but it is not the same to understand virtue and self-control as to possess them, or to understand what it means to be free of impediments as to actually be free of them. As an old Zen tale relates:

"A young student visited one master after another. Wishing to show his wisdom he called upon yet another, and said: "The mind, Buddha, and sentient beings, after all, do not exist. The true nature of phenomena is emptiness. There is no realization, no delusion, no sage, no disciple. There is no giving and nothing to be received." The master, smoking quietly, said nothing. Suddenly he whacked the student with his bamboo pipe. This made the youth quite angry. "If nothing exists," inquired the master, "where did this anger come from?""'

We often wish to be free of something or have something such that we'll imagine we do even when we do not; as Confucius observed, "all around one can see nothing pretending to be something, emptiness pretending to be fullness, penury pretending to be affluence." One can explain to another what hardship or suffering is like, but this doesn't mean they know what it is for themselves; one can spend all day discussing the difference between hot and cold but unless one has experienced it one will be none the wiser. True knowledge depends on what we discover and realize for ourselves, not just what we are told. We can't obtain virtue from another as if it were something to be handed over. "It is not much knowledge that fills and satisfies the soul, but intimate understanding of the truth," as St. Ignatius said. And this understanding doesn't just come from book-learning; it comes from realizing virtue for oneself. Theoretical insight can't replace practical experience and no amount of knowledge will help us attain wisdom if we don't put it into practice in daily life.

If taught correctly, anyone can understand truth with the intellectual mind. It is quite simple. Realizing the truth is realizing that there is not really any 'you' apart from 'It'. But knowing it isn't understanding it; words can lead the way, but they cannot open another's mind. Enlightenment is foremost an experiential transcendence of the ordinary perspective, not an intellectual one. As it is said, "One must have the face of the truth to see it. Just one such glimpse will be enough, but if you talk of "clearly seeing", you have already fallen into duality." And the only way we can assess when others have attained wisdom is from their conduct; from how they actually live. We should only be sure to judge from their real conduct and intentions and not their appearance; since some of the best people have seemed to others bad, and some of the worst have seemed to others good.[36] Outward behaviour is an indicator of inward worth, but they are not the same thing since appearances can be deceptive. One who has attained wisdom will be the last to profess it, the most ready to receive criticism and the most earnest in their spiritual pursuit. Learning and practice don't cease once some insight is attained; attaining insight is only the beginning of true learning, and as soon as one is busy talking of attaining an end or attaining a great insight or wisdom, there is no wisdom or insight because there is no practice.

Achieving wisdom

There has been an idea that attaining enlightenment comes at an instant; as a sudden and final realization. But this doesn't mean effort or training aren't required. Realization may come in an instant but the process of its attainment is long and arduous. One must bend one's life and will to this purpose: to let it permeate every thought and action. There are no secret paths or techniques to this. It doesn't lie in the pursuit of exciting sensations or mysterious experiences, or in emotional indulgence, or getting stuck on what seems clever. It lies in improving ourselves in this mundane reality; not being led by the grip of our imagination to seek something beyond it. To paraphrase a Zen Master, many students become ill because of this imagination:

> Students with illness of the eyes and ears think staring intently and nodding the ear are Zen. Those with illness of the mouth and tongue think crazy speech and loud shouting are Zen. Those with illness of the hands and feet think pacing one

way and another, pointing this way and that are Zen. Those with illness of the heart and guts think penetrating the mystery and studying the wonder are Zen.

From the perspective of reality, all of these are illnesses.

There's nothing confusing or mysterious about the true way. If you want to do something, or to attain something, or to communicate something, then do it: simply, clearly, efficiently and directly, with spontaneity and in whatever way is appropriate to the situation, without ceremony or distraction. So it's said, "do, don't just talk" and, "be mindful of the present moment." It's said by some that in awareness one should seek to see everything as new, but it isn't necessary to see everything as 'new'; it's only necessary to pay attention to things as they are. When it's said that after many years a student has attained enlightenment and is advanced enough to set out alone, this doesn't mean he has attained all there is to attain. It only means he has no need for any other teacher than himself; that he has reached a sufficient stage of understanding to seek perfection of his own accord. When this understanding is achieved one will need nothing more. But perfection is a hard thing to attain since it not only requires the attainment of right understanding, but also the ability to live in accordance with this understanding at all times and in all circumstances.

Chapter 17

Overcoming obstacles

Since the main part of achieving wisdom consists in doing, it's in everyday work and conduct that pursuing wisdom really lies.[37] We should strive to be free of all negative habits: all insincerity, laziness, conceit and self-indulgence. And we should learn to control all negative emotions like anger, envy, greed and hatred. Emotions like anxiety, fear and frustration have their uses: it's because of them we're cautious, avoid danger, and resolve to change things when stuck. But we should be cautious, diligent and strive for improvement without being anxious, fearful or frustrated. The hardest of all to master are the obstacles of ego and desires of sense. Indeed this is what pursuance ultimately consists in: in conquering these desires and being free of the attachments that spring from them. As the Katha Upanishad teaches: "Not even through deep knowledge can the atman be reached unless evil ways are abandoned, and there is rest in the senses, concentration in the mind, and peace in one's heart."

In achieving this freedom the most important step is discarding one's own ego. Desires spring from what the 'I' wants, but when there is no 'I' the only desires that govern behaviour are the ones you choose to have. Yet in order to achieve this "in many ways the soul must die to both itself and to all things it has affection for: to its love of self and its own inclinations," as St. Ignatius says. This means learning to be free from our own self-importance and our own desires; not controlled by them. The secret to achieving this is to make our attention like our desire: constant and inexhaustible. We shouldn't allow ourselves to be impeded by any-thing but instead "learn to see God in all things and seek God in all things." Virtue doesn't come from being born free from imperfections or free of obstacles. It comes from struggling to overcome them; from remaining steadfast in the face of difficulty, even when this is greatest, and even to the point of impossibility.

Fear of death is a result of both ego and the desires of the senses. It's because we have pleasure from the senses, and because the ego makes us afraid to think about the extinction of 'I' that we fear death. But the answer is not to think of birth and extinction or fear the loss of 'I', but to realize that as we pass from the eternal into the transient so we return again to eternity. We shouldn't fear death. Instead, we should be determined to make ourselves true in all things, penetrating the very heart of truth until we ourselves become it. True fearlessness is conquering our own mind and self, pursuing honesty in all circumstances, and determining to free ourselves from all delusions; giving up what we like and following what we need. This is what is hard to do. Our only concern should be to fulfill our purpose and duties; what else comes and goes is the natural way of things. Therefore, we should search after God and not fear whether we shall live or die, for death comes to all but virtue to only few.

Cultivating virtue

The meaning of life is both to think of God and to pursue God. Not merely to think of the good, but to be and do good; to strive that all hatred, greed and ignorance should be superseded and engulfed by this good. And the most important thing in pursuing this is that we should practice it ourselves: not only see the right path, but choose it and walk it. We should never think some actions matter and others don't. Everything matters. Therefore we should strive in everything we do to practice virtue; to "purify the mind in its thinking, the will in its desires, and the body in its actions." There is no single rule for moral guidance. Our intentions, actions and their consequences all matter in guiding what we do.[38] The starting point is treating others as oneself: the ancient golden rule demanded by the empathy of emotion and the justice of reason. Thus first and foremost is achieving the right balance of seeking happiness for ourselves and helping others do the same. The practice of seeing others as oneself and oneself as others is essential. In thought and action we should extend our kindness with openness and true generosity, giving without expectation of reward. It is good as we teach children to enjoy giving and receiving, but it is even better to give without attachment; to give without making any obligation in speech or thought, "to give, and serve, without any idea of self or thought of reward," as the Sufis say. As long as actions are governed by the narrow ambition of self-interest their yield will be without virtue.

We should be diligent, tolerant, take advice and work hard. We should be deferential without being accommodating, discreet whilst being open, cheerful without being mindless. We should learn courage, by which we demonstrate the difference between fearlessness and recklessness, since it's wise to stand fearlessly for what's right, but ignorance to be reckless with our own life or those of others; and whilst bravery and honour have their place in the man too ready to use them they become nothing but vanity, pride and violence. And we should be free from self-importance, which is the mark of true humility, since it's a virtue to have some self-respect but it's extreme foolishness to let this become pride or diminish into self-abasement, which are both equally the marks of an affected ego.

It's a concept amongst some that life owes us a living, but the Earth owes no-one a living: it is our duty to make one. It may not always seem possible to combine the need to work with pursuing virtue, but it isn't enough to earn a living by any means: we must do honest work. In practice it's essential that even in doing small things we should keep our eye on the bigger picture. Whatever our work is, whether it's the smallest or the greatest, we should make it our prayer and excel in doing it. We dislike doing things we feel are without purpose; but everything has its use and we shouldn't struggle against circumstances but instead try and seek out our fate in all things. In engaging the finite we should carry the infinite in mind: whilst walking on earth we should not forget about the vastness of the heavens. Not getting bogged down in everyday things we can be free to pursue virtue in our work instead of being a servant of it.

Criticism and pride

Criticism and self-criticism are the great tools of all students, and study of both oneself and others is the means by which one improves oneself. None are born free from error: the difference between the wise and foolish is that the wise seek to know and correct their errors, whilst the foolish are happy to conceal them from themselves and others and remain as they are. We should be open and content to hear the advice of others, quick to learn and slow to judge, and make use of criticism wherever it may come from. Even if we're criticised unreasonably there's no use in getting aggrieved, since anyone foolish enough to be enraged by another's criticism only proves he deserves it, for if the criticism is valid it should be accepted, and if not then what does it matter? "Useful men carrying on useful work do not become angry if they are called useless. But the useless who imagine that they are operating in a significant manner become greatly infuriated if this word is used about them," as the Sufis observe. We should learn to have an equal attitude to praise and criticism, and discard feelings of wounded pride and discouragement, and a dislike of hearing criticism or love of praise for the impediments they are. As long as we have a dislike of one or a like of the other we will be at the mercy of our own delusions and unable to see the good because of the obstruction of our own deficiencies. As it's said: "As long as you dislike it when they say you are wrong and like it when they agree with you, then good and bad cannot be distinguished."

To have self-control is to be in control of yourself in all circumstances, and to be free from pride is to be unconcerned whether you are insulted by others, treated with disdain, or eclipsed by another. It is only to be concerned to do what is right and appropriate to the situation. Indeed as a Christian master observed, "sometimes it is more difficult to endure a single word of insult, which is insignificant in itself, than a heavy blow for which we have prepared ourselves." The first prescription for one who wishes to attain wisdom is that they must learn to study and examine themselves in every instance. There are many who think they are free from self and the delusions of pride, but the moment they are insulted they leap up in indignation, and the moment they make an error they seek to justify their mistake. What can be more foolish than getting angry when insulted, or attempting to ignore one's own errors? And what can be more foolish than professing to follow God whilst only choosing to do those things one wishes to do and that path which one wishes to follow? If you are just doing what you like what has that to do with pursuing the narrow way of perfection? The right attitude is to maintain an open and devout heart at all times and be prepared to learn from anyone who can teach you; even if that is a five year old child.

Chapter 18

Dedication and expectation

No-one is able to produce a master-work at the first attempt. Only slowly by degrees does improvement come until through incessant practice one comes across the end one seeks. But defects are hard to eliminate and at every stage there's the danger of losing sight of one's goal and becoming forgetful in practice. It's no better to be in too much of a rush than to lag behind, or to seek things before one has earned them. If we expect hasty completion with little effort we will not succeed: we must be willing to persevere even when the times are hard. Hence a Sufi teacher taught, "when you feel least interested in following the Way which you have entered, this may be the time when it is most appropriate for you," and so it was said by a Zen master, "When you accumulate virtue with long practice, you do not see the good of it, but in time it will serve its function. If you abandon right and go against truth, you do not see the evil of it, but in time you will perish." We never choose difficult circumstances but it is difficult circumstances rather than easy ones that are often responsible for making us wiser. There are many who get lost on aiming at great ends because they don't first attain honesty and virtue within; still thinking of constructing tall buildings they fail to lay the foundations properly. Learning to honestly scrutinize ourselves and our actions, and patiently striving to improve our imperfections are the only means of achieving this.

We shouldn't be deceived. To pursue the true way is to walk upon a razor's edge. It is to beat down delusion, ignorance and bad habits with the strength of one's whole being. One's whole existence must be bent on striving for it; on striving to improve one's character step by step. How many are found to be truly aware, even and virtuous in all they do? Controlled, just and wise in all actions, fulfilling all in its right proportion? Freed from constraints, craving and desires, from bias and prejudice? Pride and anger all gone, humble, open and contented? It's easy to read of such an end and aspire to it, but difficult to put it into practice and through long effort achieve it. This requires unity of purpose, calm and persevering effort, and a dedicated seeking with energy and devotion.

Even when one meets with setbacks and progress seems far off, one should stick with each step by step improvement and not give up. Sometimes God seems near at hand, at other times far apart, but this isn't surprising since there are many desires and at first the mind strongly resists attempts to govern it. At times one is filled with interior joy and purity of mind; at other times plagued with restlessness and distraction, and the mind turns to what is low and destructive. As St. John of the Cross expressed, "During the night itself all hope of daylight is clouded by the darkness the soul experiences," but gradually through perseverance comes peace, tranquillity, and purity of soul; the spirit becoming "humble with respect to itself and to its neighbour, so that it is no longer disturbed and angry with itself for its own faults, nor with its neighbour for his, neither is it dissatisfied with God." We should be diligent. It's only through the experience of having one's spirit refined that one can cross to the other side. Only gradually by the cultivation of virtue and the improvement of self do one's confusion, unsettled rushing about, and lack of resoluteness become replaced by direct knowing, purposeful action, and quietude.

Living in the world

The sages say a wise man shouldn't prefer honour to insults, riches to poverty, health to sickness, a long life to a short life; he should only seek to do God's will and accord with his fate. But this doesn't mean we should disregard our health or discard honour if we find it. It just means we should use them wisely, be free of unnecessary attachments and desire nothing contrary to God's will. True freedom doesn't abandon self any more than it attaches to it. The right attitude is to live in society without being addicted to it, and to aim to excel in all we do whilst maintaining the same principle of mind in all circumstances. The material world is just as much a part of the ultimate truth as mind, but we shouldn't seek the truth either in the world or apart from it, for though the true nature of reality is expressed through phenomena and our perception of them, it's both part of them and separate from them. Those who misunderstand try to shut out phenomena without silencing their own mind; the wise silence their minds and let phenomena be as they are. "Do not permit the events of your daily lives to bind you, but never withdraw yourselves from them."[39] Live in the world and delight in it, whilst at the same time remaining detached in all things and keeping liberty of spirit in all things.

All the happy life requires is a healthy body and mind, family and friends, a roof overhead, food to eat, water to drink, freedom, peace, and as many possessions as are really required to live. Beyond that luck or fortune is incidental. "The true joy for the enlightened is in what seems ordinary to others," or as Epicurus said, "Not to be hungry, not to be thirsty, not to be cold … If someone has these things he might contend with Zeus for happiness." When one realizes the true joy of existence what concern is gain or possession? Some property is necessary but superfluities in possessions and wealth as much as those of the mind only make for distraction and lack of clarity. Simplicity of living and tastes is essential, and one should never forget that "you possess only what will not be lost in a shipwreck."

In the beginning, progress is made by making the pursuit of the true knowledge of oneself and the nature of reality one's governing desire: by subsuming all desires under the desire to achieve enlightenment and realize the truth. But afterwards one shouldn't have any such idea in mind; any thought of chasing after or seeking anything, since the harder you try and grasp it the further off it will be. By realizing true virtue within oneself one realizes something which cannot be corrupted, lost or stolen, which becomes indivisible from the self which carries it wherever it goes, and which cannot be separated from it: a good that is inseparable and without limit. The realization that "Nothing outside can cause you any disturbance." As it has been said, "Where your treasure is, there will your heart be also." True treasure is in your own mind: in appreciating existence and what it offers. The virtue of cultivating self-contentment is essentially learning to appreciate what you have: learning not to take things for granted and to make use of what you have.

We should never stop working for equality and justice: to rid the world of all exploitation and injustice, and to achieve equality and prosperity. But we shouldn't envy what others possess or covet anything. Few things are required to sustain the happy life, and all we need in this world is to be good, work hard and help those around. And if we have the misfortune to be born to a time or place where we are denied equality or justice then we should work for the attainment of these things for ourselves and others. Equanimity does not mean suffering injustice without action. It means where action is required then act; where something is beyond your control, bear it without affliction.

Realization

The last step of pursuance is that virtue so often talked of and so little seen: anger, greed, pride all gone. Content with little, free of cares, holding compassion for all, without a trace of 'I' or 'mine'. Through great struggle the self and its desires are mastered, and through great effort one has attained the effortless concentration of true wisdom. Aligned to the true principles and duties of life, with a mind fixed on God, calm and attentive, maintaining a joyful attitude, "One's house is fully at rest." Only a silent understanding remains. It is said in the Bhagavad Gita:

> The power of the senses is great, but greater than the senses is
> the mind. Greater than the mind is reason; and greater than
> reason is He: the Spirit in man and all. ... When the mind of
> the wise man is in harmony and finds rest in the Spirit within,
> all restless desires gone, then he is one in God. Then his soul
> is a lamp whose light is steady, for it burns in a shelter where
> no winds come.

No delight of the senses, no wealth, no fame, no honour or
pleasure can bring this. It requires the acquisition of no object or
possession. Only in the true knowledge and mastery of oneself,
and the realization of the nature of reality is this known. In the
end this means both achieving happiness in the world, in our
daily lives, but also achieving some understanding of the
ultimate nature of things. "Truth is more than mere happiness.
The man who has truth can have whatever mood he wishes, or
none."

It is said by the masters "Before enlightenment I ate and slept;
after enlightenment I ate and slept." But the meaning isn't that
there is no difference: the meaning is the same things are done
afterwards but they aren't the same. "I went away from home
empty-handed, and empty-handed I returned." Yet they aren't
empty, and you see that when you thought you were alive, you
were not at all; before, the mind was negligent, afterwards the
mind is mind-full. What seemed ordinary becomes extraordinary and the dull and mundane glitters with a silent and
inexpressible joy. Yet as Bodhidharma observed: "Everyone
wants to see this Mind, and those who move their hands and feet
by its light are as many as the grains of sand along the Ganges,
but when you ask them, they can't explain it. They're like
puppets. It's theirs to use. Why don't they see it?" The reason
they don't see it is because they don't know themselves: everyone

walks about in the great realm, but few realize it.

The secret of wisdom is to stop holding onto things. Don't hold on to anything in your mind: neither anxiety nor excitement, desire nor anger. Neither self nor other. Let go of conditions and prejudices, of desires and attachments, of clinging to words and things. Return to the unity you were before you were born into multiplicity: enter the state of unknowing, then be as one and know all things as one. As Bodhidharma said, *"Once attached, you're unaware; but once you see your own nature, the entire scriptures become so much prose. Their thousands of verses only amount to a clear mind. Understanding comes in midsentence. What good are doctrines?"* Don't think of sitting mindlessly; become aware. At this instant, become miraculously aware, in meditation and in action. Become appreciative of the present moment, cultivate your virtue in pursuit of the right path, and see where your true self lies. But don't let yourself become caught up with delusion or imagination; and don't conjure up your own illusions.

We should be attentive and become aware to this: transcendental and ordinary reality are one. There are not two realities. Absolute and ordinary realities are the same thing: here and now. Then we can see that our mortal nature is not different from our real nature. God is the essence of our existence; beyond this existence there's no God and beyond God there's no existence. *'That thou art'; not only in other people but in the universe itself.* Realize that your real Self lies within It: within this whole vast impenetrable reality that is the nature of every single form but no particular form Itself. As Guru Nanak said, "God is both the fish and the fisherman, the water and the net, the float of the net and the swimming of the fish within it." When one realizes this; the nature of this reality, one will be like the raindrop falling from the cloud that perceives the sea. There is no confusion remaining. This is what is meant by knowing yourself, by seeing that form and differentiation are empty, and by having a mind that abides nowhere.

The essence of the true way is living life just as it should be lived; utterly direct and natural. Not doing as one pleases without constraint, but through long practice being free from constraint and free from delusion, internally and externally. Humble and contented, one's actions are free from thought of attachment or reward; not wasting a moment, pursuing truth and doing good for its own sake, and having our joy in everyday life. "When mortals are alive, they're anxious about death. When they're full, they worry about hunger. Theirs is the great uncertainty. But sages don't dwell on the past. And they don't worry about the future. Nor do they cling to the present. And from moment to moment they follow the Way."[40] Or as Epicurus said: "Don't fear God. Don't fear death. What is good is easy to find. What causes suffering is easy to bear." The perfection of man is in his conduct, and virtue is the first and last step of pursuance. If we're fortunate to find one who's attained it we will know because we will see it expressed in every aspect; brilliant and sincere in all things. If we wish to attain this every-minute awareness for ourselves we should practice inside and out with all our being until our thoughts and actions become single-pointed in pursuit of this truth.

So do we realize creation doesn't just exist for enjoyment but that life is both a great responsibility and a great opportunity, and that the energy that moves our hearts and minds is the same as that which moves the entire universe. Life is a gift of infinite rarity and nothing of worth is achieved in it without effort. We can always lose hope for the world seeing things as they are, but we should not lose faith. It is in our power to make of the world and ourselves what we will. Just as we must work internally for enlightenment, so we must work externally to earn our living and contribute to the world. We are fortunate to live in a time when our freedom has been much enhanced: our rights, knowledge and understanding no longer so confined by custom, superstition and inequality, but freed by the effort and struggle of those who have gone before. Yet there remains much to be done. Life and death matter. Our mundane existence constitutes a part of the whole process of life: of the thoughts, events and actions that are the universe. So we should be sure that we take part and make our contribution in whatever sphere it may be. We should let our pursuit of the spiritual be the well-spring from which all else flows and let it permeate our life in every way. Learning to keep to the right path in all circumstances we can attain true happiness for ourselves and for others.

This one moment is an infinite amount of time
An infinite amount of time is at the same time this one moment
If you see into this
You will realize the self which is seeing it

Wumen Huikai

Notes

1   It is widely held that science is a separate field to philosophy and has replaced 'natural philosophy' in understanding the physical universe; but a better way to look at it might be to see that natural philosophy has just improved its methods and equipment.

2   'Metaphysical' meaning any effort to describe the ultimate nature of things.

3   Indeed the very foundations of most life, chloroplasts and mitochondria, are symbiotically derived.

4   There is also a third problem: which is that even a proof of miracles occurring would not necessarily prove the existence of any God, e.g. if said purported miracle is a statue weeping whilst thousands are dying of the plague in the next village, one might question whether the miracle is of help in proving the existence of an omnipotent benevolent creator.

5   In Epicurus words, quoted by Hume: "Is God willing to prevent evil, but not able? Then he is not omnipotent. Is he able but not willing? Then he is malevolent. Is he both able and willing? Then whence evil?"

6   It has been argued that limiting human freedom to cause harm to others should be feasible: but how can you feasibly limit freedom in this respect without also limiting it in a thousand other ways? It is the small amount of true freedom we have that makes life meaningful to possess: our ability to not just experience but to desire ends and act for them.

7   Innate truths are ideas or concepts believed to be implanted/hardwired in the mind at birth, not learned from experience.

8   i.e. inductively: by inferring a general rule or 'law' from observing actual, particular examples.

9   e.g. when we see a 'table' all we perceive is a bunch of properties we bundle together as 'table': we see a rectangular shape of the colour brown, which appears to reflect light, and which we touch and feels hard and which we hear makes a certain kind of rapping noise, and this is what we call 'table'. We don't know what underlies or creates these properties.

10 i.e. unlike deductive reasoning (logically deducing particulars from general premises), inductive reasoning (inferring a rule or premise from observing particular examples) does not provide logical certainty: you merely observe such and such things occur and try to infer which rule is operating. But until you know the premise or rule with certainty and can reason deductively to specifics, you are confined to inductive reasoning, which has no certainty. All knowledge of the world is ultimately inductive.

11 e.g. why two chemicals react at certain temperatures or in certain combinations because of their atomic properties (and so should continue to do so in future) etc.

12 Philosophically known as the 'dilemma of determinism': the dilemma being free will is impossible because either causal determinism is true, in which case we aren't free, or not true, in which case things are random and we still aren't free. But the second option is clearly irrelevant as an alternative: human actions aren't totally random. And the problem with the first option is ambiguity in using the word 'cause' and what is meant by it. At crux is whether humans can genuinely originate actions; but clearly innumerable causes contribute to human actions. If the 'determinist' acknowledges the will itself as one of the causal factors the views aren't irreconcilable, since both determinists/indeterminists mean the will is one of the determining elements. If the determinist doesn't acknowledge will as an independent causal factor, then he either views humans as automatons

(i.e. computers responding to inputs), which is a caricature, or he views the will as itself completely determined by preceding factors. And the answer to the latter is that it is as unverifiable as it is irrefutable, useless as a guide to practical action, and morally nihilistic.

13 e.g. from theist to deist, dualist to monist, idealist to materialist, transcendent to immanent, etc.

14 Indeed, as Schopenhauer noted: "In fact it might be asserted that some absolute inconsistencies and contradictions, some actual absurdities, are an essential ingredient of a complete religion; for these are just the stamp of its allegorical nature and the only suitable way of making the ordinary mind feel what would be incomprehensible to it".

15 There have been numerous historical attempts to prove God's existence but none have withstood the test of time or scrutiny. In brief these include: (1) The ontological proof, arguing God must exist because we can conceive of him as perfect, and being perfect he must possess existence, else he would be imperfect. The short answer of course is it's not enough we can just think of God to prove he exists. A stronger argument is that we have ideas like 'perfection' which could only have come from God; but again, we have no experience of 'infinity' and can extrapolate to imagine it from our finite experience without needing God-given innate ideas. (2) The cosmological proof, a 'first-cause' argument that causes in the world cannot be infinite, and all things must be caused by something: by a first and necessary cause. The problem is we have little reason to propose one cause exists over another. (3) The argument there is such design and purpose to the world that it must have had a designer; the reply is evolution somewhat undermines the case, and even if accepted it's hard to explain why the world is lacking in so many perfections. (4) A fourth argument (Kant's) is that 'moral law' demands justice, and since this is

evidently not the case in this life where many virtuous people suffer and many bad people prosper, there must be a God and future life where virtue is rewarded. Needless to say, this is optimistic.

16 The phrase is Schopenhauer's.

17 Any particular action itself may of course turn out morally worthy or not irrespective of motives. One may also choose to credit someone with an unconscious desire to do good, on occasion.

18 i.e. a qualified, holistic form of consequentialism, or more specifically, utilitarianism, that has happiness rather than pleasure as the key principle, and doesn't commit the fallacy of assuming happiness can be achieved without consideration for means (utilitarianism being the 'Greatest Happiness Principle', i.e. the guide to moral conduct should be doing whatever increases the sum human total of pleasure/reduces pain). One of the most common complaints against consequentialist attitudes is they permit evil to be justified in order to facilitate a greater good, but situations where evil is advocated for the 'greater good' rarely facilitate the greater good at all, and a proper utilitarian approach will always regard the means as integral to any end it aims to achieve. It will also acknowledge there are principles of conduct we should follow as a rule (and hence the validity of deontology as a concept).

19 Though (per moral philosophy) this doesn't necessarily imply an antirealist view of morality (and it certainly does not imply a relativist view, which as a practical concept is useless and inconsistent, since it can advocate nothing). The debate between realism and antirealism in meta-ethics is besides largely academic: it arises because people are concerned to justify an authority, and its source, in morality: but whether you think moral law was something objectively created with our existence that we discover, or something

that humanity creates doesn't really matter. In either case we'd need to uncover what it was and judge whether our discoveries seemed true in the process.

20 The fact that we sometimes do something because we think it is the 'right thing to do' even though we don't particularly want to do it, doesn't show that reason is the true driver of objective moral judgement. Reason assists the process, but it still ultimately comes down to making a choice about what we 'desire' to do; about choosing between desires.

21 Per Kant's rational reformulations of the ancient 'Golden Rule' (which is: 'Do as you would be done by') as forms of his categorical imperative: do as you would legislate all others to do, and treat them as an end not a means (i.e. treat others as you would wish to be treated yourself).

22 i.e. periodic general elections, an independent judiciary and executive, equality before the law, a diverse and independent media, basic consumer/worker protections, a minimal welfare support etc.

23 It's been suggested we'd be better off free of private property and if all business was run by the people for the people, not the few. But whilst we may hope a communal government would be run by the people for the people, governments can't function by committees alone; leadership is required and hierarchies form themselves. Wherever one group is given the opportunity to control others it's plain what the effects will be. If government is allowed to control all commerce and distribution of wealth it will result in author-itarianism and lack of freedom at best, and at worst exploitation and tyranny. The middle way is neither for the world to be controlled and exploited by concentrated private capital, nor for all property and wealth to be seized and controlled by the government. It is for individual property to be respected but for democratic governments to admin-ister societies for the benefit and service of all, and to tax

individuals and businesses within reason and in proportion to income (and inheritance), such that all in society may be provided with the necessities; whilst also ensuring that such assistance isn't exploited and people learn to help themselves rather than relying on others to do so. This doesn't mean business should be lawless; it means the market should be governed by laws that justly protect and preserve the rights of all, balancing freedom and efficiency with prevention of unfair practice, monopoly, dishonesty and negligence. Total freedom isn't free at all: it is anarchy. True freedom requires just law. The dangers of unregulated private commerce are greed and exploitation dominate them; the dangers of public control are bureaucratic inefficiency and corruption. What a free society needs is a free market, but a regulated free market and a government able to strike the balance in regulation. There will always be difficulty reconciling the reward of hard work and right to property with achieving a completely even distribution of wealth. But this is no great problem so long as all are provided with the necessities of life, for wealth broadly in proportion to effort and initiative is largely as it should be.

24   It is no coincidence that science, our most profound and true knowledge of the world, has only flourished in those regions that have established a stable and genuine democracy; because democratic characteristics, such as transparency, accountability, and freedom of speech and critical debate, are also essential for good science.

25   In the words of J.S. Mill: "Better [i.e. happier] to be a human being dissatisfied than a pig satisfied; better to be Socrates dissatisfied than a fool satisfied."

26   As Popper observed, this burden of personal responsibility has been increased by "the strain of modern civilization"; by the breakdown of the certain structure and close support of the tribe and its replacement with the emotional distance of

the vast and anonymous modern society. But only an illusion of nostalgia would make us wish to return to an age of superstition, inequality and illness; to chase an imaginary paradise that mostly did not exist and cannot be recreated. Just like the burden of personal responsibility that wisdom brings, the burden of the modern society is a healthy one: it demands we learn, work and take responsibility for ourselves and make our own contribution to society.

27 A somewhat analogous situation exists in commerce where it is considered 'good business' to make money by minimising costs and maximising profits, and indeed there is no use trying to produce something if it costs more than it yields. But when a business is run as if this is the sole criterion of success there is a great deal wrong because one will naturally come to regard all employees, partners and customers as part of the cost equation and regard them not as people but only as a means or hindrance to making money. The life of man is brief, it does not require a great deal to attain happiness whilst it exists, and when it ends it takes nothing with it. Any good businessman should see himself as an integrated part of the community, not apart from it and prepared to exploit profit from it by any means. Moreover, it is bad business to operate with profit as one's sole aim, since exploitative leadership in business as in politics, only breeds resentment and enmity, and malpractice and dishonesty are certain to undermine one's enterprise whether sooner or later.

28 Bodhidharma.

29 The meaning of this quote is that there is no virtue in asceticism if it leads to bitter misanthropy: better to wear fine clothes, eat well, and be full of love for others.

30 How great is the irony that politics, the most desperate cause of mankind, which needs the most able and least self-interested characters, often attracts the opposite: dishonest

people seeking their own particular good, whilst the good men of the world disdain to engage in the servile struggle to obtain popular favour and power. But whilst it may be thought those who become caught up in politics will be forced to sacrifice their principles, in reality the pretence that 'good sometimes requires evil' is the most overused and rarely justified excuse offered by those who wish to disguise the choice they have made as the better, when in fact it is the worse. It has been a favourite pastime of men to practice tyranny and call it freedom, but we should recognize that we should not take at face value what we are told or what others would have us believe. How often does one see vice dress itself in the propaganda of virtue?

31 Diderot's c.1749 letter to Voltaire was not the first to express this sentiment.

32 The phrase (in Latin: sub specie aeternitatis), is Spinoza's.

33 The common aim of meditation is to still and concentrate the mind and achieve a 'one-pointedness' of concentration by setting aside distractions and involvements. With an attentive concentration, without tension, one attempts to discard the mind's wandering and 'be attentive to what you are doing', until one is able to forsake all thoughts of self, and all thoughts of sense and desire, past and future, and focus the mind without intrusion. By this folding in of the mind one aims to achieve a kind of introversion without being introverted: a true *hesychia* or stillness. When one can enter into this meditation without any discontinuation, with complete concentration and awareness, no distractions or thoughts arising, then it is said dhyana may be attained (see glossary).

34 Pseudo-Dionysius.

35 Ibn Arabi.

36 Epicurus is a case in point: famed through the ages for allegedly advocating a luxuriant and immoral lifestyle, he

actually lived a simplistic life of the highest virtue and considered the occasional pot of cheese an indulgent luxury.

37 There is an old Christian argument about whether salvation relies on faith alone or goods works: the answer is of course both. Faith is in good work; good work shows faith. As James 2:26 says: "As the body without the spirit is dead, so faith without works is dead also."

38 As J.S. Mill said, "It is not the fault of any creed, but of the complicated nature of human affairs, that rules of conduct cannot be so framed as to require no exceptions." In philosophical parlance, teleology (judging actions by consequences/outcomes), deontology (judging actions by their observation of duties, rules or principles) and virtue ethics (judging actions with reference to character) are all valid and necessary approaches to considering moral behaviour and what we ought to do. And since all depend on inductive material for refinement, all are subject to pragmatic improvement.

39 Huang Po.

40 Bodhidharma.

Glossary

All dates are CE unless otherwise indicated. Titles of texts are in italics.

Aquinas, St. Thomas (1225-1274): An Italian Dominican priest and amongst the most influential philosopher-theologians of Christianity. Aquinas' blend of Greek philosophy and Christianity cemented Roman Catholicism in affirming reason alongside revelation as a means to access and judge the highest truth. He gave up a life of aristocracy for a religious career as a scholar of increasing authority, though his life concluded with a spiritual experience that led him to state "all that I have written seems like straw." Much of his life was spent teaching between Paris and Naples.

Ibn Arabi (1165-1240): An Andalusian Moor and amongst the most prominent of Sufi mystics. Known as Sheikh al-Akbah ('The Great Master'), Arabi claimed himself a disciple of Khidr (the mysterious figure alluded to in the *Quran* 18:69) in his pursuit of transcendent truth. His output was prolific in Islamic theology and philosophy. After studying in Andalusia/North Africa he left for Mecca in his early 30s, travelling extensively before settling in Damascus c.1220.

Atman: Sanskrit for 'self', typically referring to the true self that is identical with Brahman or Paramatman, the infinite and universal Spirit that is the foundation of the universe. By acquiring self-knowledge, the individual realizes his true nature is synonymous with this ultimate reality beyond the veneer of the phenomenal, material world. 'A drop of water in water' is one of many poetic analogies applied to the monistic or pantheistic Atma/Paramatma relationship. See 'That Thou Art'.

Aristotle (384–322 BCE): Student of Plato and teacher to Alexander the Great, and a figure of almost equal significance to his instructor in Western Philosophy (and of greater impact on Islamic philosophy). Aristotle's empiricism is traditionally contrasted with Plato's rationalism, whilst his breadth and systematisation of philosophical thought was unparalleled in scope and impact. Barring a decade teaching in Macedonia, Aristotle spent most of his adult life in Athens where he founded the Lyceum.

St. Augustine (354-430): Bishop of Hippo Regius (in modern-day Algeria) and arguably the most important philosopher-theologian in Christianity. Augustine formalised Christian incorporation of key elements of Greek philosophy, was one of the founding doctrinal fathers of the Catholic Church, and made significant contributions to secular philosophy. A few years in Rome and Milan aside, Augustine's life was spent in North Africa where he founded a monastic house c.391.

Bhagavad Gita: A short scripture that forms part of the poetic Sanskrit epic *Mahabharata*, which is dated to the c.eighth-second centuries BCE and is one of ancient India's two great classics (the *Ramayana* is the other). The *Gita* consists of a dialogue between a warrior prince, Arjuna, faced with the horror of war, and his divine spiritual guide Krishna, and is considered complementary to the *Upanishads* in Hindu philosophy (and probably of greater significance to most lay practitioners).

Bodhidharma (fifth/sixth century): A Buddhist monk of Indian or Persian origin, credited with introducing Zen Buddhism to China. After arriving in China Bodhidharma is said to have spent nine years in meditation facing a cave wall close the Shaolin monastery. Zen practitioners regard him as the first (of six) Patriarchs, whilst legend also credits him with training Shaolin monks in yogic exercise that led to the creation of Kung-Fu. Custom depicts Bodhidharma as a 'bearded barbarian' minus eyelids (allegedly torn off to prevent himself falling asleep during said meditation).

Confucius/Kongzi (551-479 BCE): Eponymous founder of one of China's most influential philosophical traditions, Confucianism; a humanistic system emphasising personal morality and the importance of social relationships and family respect. Confucianism's order and hierarchical outlook is often contrasted with Taoism's dynamic fluidity, though the two are more complementary than opposing. Confucius was a teacher and statesman of Lu province.

Dhyana (Sanskrit, in Chinese: Chan, in Japanese: Zen): The term for the meditative practice of Hinduism/Buddhism by which the practitioner aims to achieve Samadhi, a state of detached, absorbed and undisturbed consciousness that leads to transcendence of duality and 'enlightenment'. Whilst most Eastern traditions outline such meditation as the ultimate end of religious practice, it has become most famously associated with the Zen branch of Buddhism.

Meister Eckhart (1260-1328): A German Dominican theologian. Eckhart's free-thinking mysticism brought him into conflict with Church authorities on more than one occasion, culminating in some of his views' eventual condemnation for heresy by Pope John XXII in 1329 (Eckhart disappeared, presumed deceased, 1327/8). Highly educated (largely at Paris), Eckhart drew on influences from Plotinus to Aquinas, yet endeavoured to make his inspirational sermons as accessible as possible.

Epicurus (341-270 BCE): An ancient Greek philosopher famous for his ethical view that the purpose of life was to achieve happiness by maximising pleasure and minimising pain, his empirical outlook and rejection of a providential theology (or afterlife), and his atomic theory of the natural world. Misrepresented through history as an infamous hedonist, Epicurus in fact led a life of extremely disciplined simplicity and advocated a responsible life free from fear and anxiety.

Gilgamesh: An ancient Mesopotamian (modern-day Iraqi) poem dated to the early third millennium BCE and amongst the earliest extant works of world literature. Besides its recount of familiar human themes and its age, *Gilgamesh* is notable for predating key biblical themes with its account of a flood myth and tale of the loss of tribal innocence (per Adam and Eve).

Hesychia: From the Greek for silence, rest or stillness. Hesychasm is an Eastern Orthodox Christian eremitic prayer practice that aims to achieve a direct experiential knowledge of God, typically by use of mantra (e.g. the Jesus Prayer), initially vocally and subsequently as silent meditation. With its aim of Theosis (union with God) Hesychasm is highly reminiscent of Eastern supraconceptual meditation techniques. Its use reflects the Eastern Orthodox Church's early divergence from the Roman Catholic Church in relegating reasoned philosophy as subordinate to the role of mystery, contemplation, and silent meditation in achieving knowledge of God. This divergence long predated the formal split between the Churches in 1204 (after the Crusader's infamous sack of Constantinople), and has largely continued into the modern era (e.g. the Roman Catholic Church's seventeenth century outlawing of Quietism as heretical).

Huangbo/Huang Po (deceased c.850): A famous Zen Buddhist master, popularly named after the mountain with which he was associated as a monk (Mt. Huangbo in Fujian). His two key known texts were posthumously transcribed by a student, P'ei

Hsiu. Huang Po was master to Linji, founder of the great Linji (Japanese: Rinzai) Zen sect.

Huikai, Wumen (Japanese: Mumon Ekai) (1183-1260): A Chinese Zen master known for his compilation of the famous koan-collection, *The Gateless Barrier*. A **koan** is a meditation device developed by some Zen schools to challenge students to transcend their ordinary way of thinking (i.e. logical, dualistic conceptualising). Typically a short puzzling statement or story, a student may end up working with a single koan for years (i.e. effectively as a prayer mantra).

Hume, David (1711-1776): A key figure in the Scottish Enlightenment and amongst the most important philosophers in Western History. Hume was third of the great British empicirists (after Locke and Berkeley), demolishing many religious and rationalist assumptions with definitive arguments relating to induction and the importance of skepticism in epistemology. Above all philosophers Hume did most to define the limits of human knowledge, a situation that has arguably never been altered (despite Kant's attempt). Barring some years in France, much of his life was spent in Edinburgh.

St. Ignatius of Loyola (1491-1556): A Spanish nobleman severely injured in the battle of Pamplona (1521); Ignatius underwent spiritual conversion during convalescence, inspired by *De Vita Christi* (amongst other works) to pursue a religious life. He authored the famous *Spiritual Exercises* (1522-1524) and, after studying in Paris for seven years, emerged as a key figure in the counter-reformation with his 1534 foundation of the Catholic missionary religious order, the Society of Jesus (Jesuits).

St. John of the Cross (1542-1591): A Christian Spanish mystic and author of the famous poem *Dark Night of the Soul*. Born into poverty, St. John studied as a Jesuit before entering the Carmelite order in 1563, where he joined forces with another famous mystic, St. Teresa of Avila, to found the reformed order of Carmelites (the shoeless or 'discalced' Carmelites). The *Dark Night* was started whilst St. John was imprisoned in solitary during 1578 by his fellow Carmelitan brothers, unhappy over his efforts to reform the order (presumably the prospect of going barefoot did not appeal).

Kant, Immanuel (1724-1804): A towering figure in Western philosophy, famed for a metaphysics that turned received wisdom on its head by arguing that time, space and causality are integral aspects of our apparatus of perception rather than what we perceive (formally: space and time are *a priori* intuitions of sense perception, and causality is an *a priori* concept of the understanding). Kant's key contribution was the realization that the human is not merely a receptacle for information, but that we bring a significant influence to bear on our interaction with reality. In morality, Kant's key contribution was an intensely rational approach to formulating a universal guide of moral conduct. His entire life was spent in Prussia.

Laozi/Lao Tzu: Chinese for 'Old Master' and the apocryphal author of Taoism/Daoism's foundational text, the *Tao Te Ching* (sometimes simply known as *Laozi*). A short and esoteric text, *Laozi* emphasizes individual philosophical/religious practice focused on living in accordance with the Tao (see 'The Way'). The text's ambiguity has lent itself to much historical misinterpretation (e.g. in support of anarchism). Alongside Zhuangzi's *Chuang Tzu*, the book forms the short scriptural canon of Taoism. It is traditionally dated to the sixth century BCE.

Leibniz (1646-1716): A German mathematician and philosopher known for significantly advancing the world's first mechanical calculators and credited with co-discovery (alongside Newton) of calculus. Leibniz is perhaps best known for his argument that "this is the best of all possible worlds."

Locke, John (1632-1704): A British philosopher of pre-eminence during the European enlightenment whose sane and rational liberalism were of profound importance. Locke's empiricism paved the way for Hume and contributed an understanding of the importance of experience in shaping human knowledge and character (e.g. as in education), whilst his political philosophy had a primary influence on the U.S. declaration of Independence (Jefferson called him "one of the three greatest men that have ever lived").

Mill, John Stuart (1806-1873): An influential British philosopher and subject of a famously intense education (including learning Greek at three and Latin at eight) that precipitated a nervous breakdown age twenty, from which he recovered with the aid of poetry. The training, devised by his father and Jeremy Bentham, did however succeed in producing an intellect capable of furthering the cause of utilitarianism. Mill's most significant other contributions were in political philosophy in support of democracy, liberty, and the rights of the marginalised and oppressed.

Guru Nanak (1469-1539): Founder of the Sikh religion and first of ten Sikh Gurus, Guru Nanak was a legendarily well-travelled Indian mystic born close to Lahore (modern-day Pakistan). Apparently seeking to transcend religious corruption and the doctrinal/political disputes amongst contemporaries, Guru Nanak endeavoured to pursue and preach a spiritual path that transcended such differences, which became the eventual basis for the Sikh religion. **Sikhism** advocates equality amongst all peoples and between the sexes (including rejection of the caste system), a minimum of ritual and intermediation (there are no priests), a simple monotheistic worship, and a practical ethical purism whilst living as a responsible householder (encapsulated by the mantra: vand chakko (share with others), kirat karo (earn an honest living), naam japo (worship God)).

Neng, Hui (638-713): A Chinese Zen master and allegedly illiterate peasant whose profound intuitive insight saw him become the sixth and final patriarch of Zen Buddhism. Hui Neng is generally attributed with original authorship of the *Platform Sutra*, the only Chinese Buddhist text accorded sutra (scriptural) status, and is the subject of a famous tale about the succession to the fifth patriarch.

Plato (c.427–347 BCE): Student of Socrates and most prominent of the Greek philosophers, known for his depth and beauty of philosophical thought and the Socratic dialogues by which the latter's teachings are known. Plato is most famous for his theory of Forms: the view all particular things we perceive have their form from an archetype Form that exists in a changeless, eternal and perfect realm only perceivable by the mind. It has been claimed, not totally without justification, that all of Western Philosophy is but a footnote to Plato. His life was spent in Athens where he established his famous Academy.

Popper, Karl (1902-1994): Most famous for the falsification concept in the philosophy of science and his arguments for democracy in political philosophy. Falsification theory stood in contrast to then-prevailing ideas such as logical positivism/verificationism, and illustrated that historical materialism (e.g. Marx) was not a science, whilst arguments for democracy exposed the fallacies behind totalitarianist politics. Born in Vienna, Popper moved to London in 1946.

Pseudo-Dionysius/Pseudo-Denys (late fifth-early sixth century): An early Christian philosopher-theologian whose works were in antiquity incorrectly attributed to St. Dionysius the Areopagite until recent scholarship revealed them to be the work of an unknown fifth century author, now named 'Pseudo-Dionysius'. His writings fuse mystical and Neoplatonic influences and had a lasting influence on Christianity, and were popularised by Johannes Scotus Eriugena's ninth century Latin translations.

Rumi, Jalal ad-Din Balkhi (1207-1273): Author of the *Masnavi/Mathnawi*, known as the 'Persian Quran', and one of Sufism's greatest poets and exponents. Rumi was born in Balkh (modern day Afghanistan/Tajikistan) before fleeing to Turkey in advance of Genghis Khan. Already influenced by persian poetry and his father's Sufism, a 1244 meeting with Shams-e Tabriz, the wandering mystic who briefly became Rumi's spiritual instructor, was (according to legend) transformational; inspiring the subsequently recited *Masnavi*. Rumi's followers posthumously founded the Mevlevi Sufi order, known as 'The Whirling Dervishes'.

Schopenhauer, Arthur (1788-1860): A German philosopher famous for his post-Kantian philosophy infused with elements of Eastern mysticism. Frequently portrayed as a bleak pessimist, this arguably reflected Schopenhauer's genuine understanding of man's predicament and state of the natural world, his conclusion that conscious existence beyond death is not probable, and his honest assessment of the conduct of most of his fellow men. His most underrated insight is perhaps the observation that Kant had missed one thing we can truly know 'in itself' (i.e. without intermediation of the senses), which is ourselves, and this knowledge of self was the key to understanding the nature of reality.

Spinoza, Benedictus de (Latin, in Hebrew: Baruch Spinoza) (1632-1677): A Sephardic Dutch Jew and one of Europe's most remarkable philosophers. Excommunicated at 23 from the Jewish community for his unorthodox views, which included a critical appraisal of religious authorities, denying the afterlife, and rejection of the existence of a providential theistic God (which for his time made him practically an atheist). Spinoza led an epicurean lifestyle of sociable and frugal simplicity, making a living grinding optical lenses. His posthumously published *Ethics,* consists of a classical rationalist attempt to *a priori* deduce the nature of God and reality (an unproven beautiful monistic panentheism) and a practical ethics.

Sufism/tasawwuf: The inner, mystical branch of Islam. Historically Sufism was closely associated with the eremitic, ascetic lifestyle, but despite the tension of its mystical outlook with scholastic Islam, its teaching and attitudes arguably became the dominant mood of Islam through the middle ages (a situation that has largely reversed). Stressing personal experience and the approach of the heart in realizing a transcendental truth beyond the reach of rational philosophy, its core teaching is to realize the annihilation of self and union with the divine by a path of self-effacement and purification. It originated as a formal movement in Persia toward the end of the first millenium C.E.

'The Way': An expression that originated in Taoism, but was adopted conceptually into geographically associated philosophies/religions such as Confucianism and Buddhism. The phrase originates in the *Tao Te Ching* as the Chinese term Tao or Dao, literally meaning way, path or principle, and is the central concept of Taoism.

'That Thou Art' (Sanskrit: Tat Twam Asi): Originally quoted in the *Chandogya Upanishad*, the phrase may also be translated as 'You are that' and encapsulates the core of the Vedanta teaching (See *Upanishads*). It is typically interpreted as referring to the unity of the individual soul (Atman) with the universal soul/nature of reality (Paramatman/Brahman).

Upanishads: Hindu scriptures whose core transcription is dated to Northern India tenth -sixth centuries BCE. *Upanishads* means 'sitting at the feet of [a master]' and refers to their orally transmitted tradition. Vedanta ('end of the *Vedas*') is the most influential of Hindu philosophy's six orthodox schools, and holds the *Upanishads* to conclude and interpret the older *Vedas* scriptures, effectively superseding their polytheistic rituality with a simple ethical monism. The **Vedas** themselves are the oldest texts of Sanskrit and Hinduism, regarded as revealed scriptural authority by orthodox Hindus. The corpus of Vedic texts are typically dated to c.1500-600 BCE.

Utilitarianism: The 'Greatest Happiness Principle' theory of ethics, advocating that the overarching guide to moral actions should be the sum total of all happiness of all human beings (or sometimes all sentient beings). Thus the aim of all moral conduct should be to reduce pain/suffering and increase pleasure/happiness. The theory is a form of **consequentialism** in considering effects (i.e. consequences) of actions to be their most important measure, a view contrasted with **deontology**, which judges actions by whether they follow certain rules, duties or principles (i.e. theoretically irrespective of outcome).

Zen (Japanese, Chinese: Chan): The mystical branch of Mahayana Buddhism that originated with Bodhidharma in sixth century China (arriving in Japan c.twelfth century). Zen places an early and immediate emphasis on meditation to achieve personal insight and experience of enlightenment under the tuition of an accomplished master, paying relatively scant attention to scholastic formality. In giving less of a role to the doctrinal and ritualistic aspects of religious practice and greater prominence to achieving direct, individual understanding, Zen achieved renown for its simplicity and innovative approach to 'expressing the inexpressible', with its use of paradox, contradictory language and practical action. While commonly misconstrued as dispensing with the requirements of formal Buddhism, Zen rather takes for granted mastery of the first seven steps of Buddhism's 'eightfold path' (which essentially relate to right conduct) and proceeds to emphasize the importance of the eighth: right concentration (i.e. meditation).

BOOKS

Iff Books is interested in ideas and reasoning. It publishes
material on science, philosophy and law. Iff Books aims to work
with authors and titles that augment our understanding of the
human condition, society and civilisation, and the world or
universe in which we live.